REGULATORY CONTROL
OF EXPOSURE
DUE TO RADIONUCLIDES
IN BUILDING MATERIALS AND
CONSTRUCTION MATERIALS

The following States are Members of the International Atomic Energy Agency:

AFGHANISTAN
ALBANIA
ALGERIA
ANGOLA
ANTIGUA AND BARBUDA
ARGENTINA
ARMENIA
AUSTRALIA
AUSTRIA
AZERBAIJAN
BAHAMAS
BAHRAIN
BANGLADESH
BARBADOS
BELARUS
BELGIUM
BELIZE
BENIN
BOLIVIA, PLURINATIONAL
    STATE OF
BOSNIA AND HERZEGOVINA
BOTSWANA
BRAZIL
BRUNEI DARUSSALAM
BULGARIA
BURKINA FASO
BURUNDI
CAMBODIA
CAMEROON
CANADA
CENTRAL AFRICAN
    REPUBLIC
CHAD
CHILE
CHINA
COLOMBIA
COMOROS
CONGO
COSTA RICA
CÔTE D'IVOIRE
CROATIA
CUBA
CYPRUS
CZECH REPUBLIC
DEMOCRATIC REPUBLIC
    OF THE CONGO
DENMARK
DJIBOUTI
DOMINICA
DOMINICAN REPUBLIC
ECUADOR
EGYPT
EL SALVADOR
ERITREA
ESTONIA
ESWATINI
ETHIOPIA
FIJI
FINLAND
FRANCE
GABON
GEORGIA

GERMANY
GHANA
GREECE
GRENADA
GUATEMALA
GUYANA
HAITI
HOLY SEE
HONDURAS
HUNGARY
ICELAND
INDIA
INDONESIA
IRAN, ISLAMIC REPUBLIC OF
IRAQ
IRELAND
ISRAEL
ITALY
JAMAICA
JAPAN
JORDAN
KAZAKHSTAN
KENYA
KOREA, REPUBLIC OF
KUWAIT
KYRGYZSTAN
LAO PEOPLE'S DEMOCRATIC
    REPUBLIC
LATVIA
LEBANON
LESOTHO
LIBERIA
LIBYA
LIECHTENSTEIN
LITHUANIA
LUXEMBOURG
MADAGASCAR
MALAWI
MALAYSIA
MALI
MALTA
MARSHALL ISLANDS
MAURITANIA
MAURITIUS
MEXICO
MONACO
MONGOLIA
MONTENEGRO
MOROCCO
MOZAMBIQUE
MYANMAR
NAMIBIA
NEPAL
NETHERLANDS
NEW ZEALAND
NICARAGUA
NIGER
NIGERIA
NORTH MACEDONIA
NORWAY
OMAN
PAKISTAN

PALAU
PANAMA
PAPUA NEW GUINEA
PARAGUAY
PERU
PHILIPPINES
POLAND
PORTUGAL
QATAR
REPUBLIC OF MOLDOVA
ROMANIA
RUSSIAN FEDERATION
RWANDA
SAINT KITTS AND NEVIS
SAINT LUCIA
SAINT VINCENT AND
    THE GRENADINES
SAMOA
SAN MARINO
SAUDI ARABIA
SENEGAL
SERBIA
SEYCHELLES
SIERRA LEONE
SINGAPORE
SLOVAKIA
SLOVENIA
SOUTH AFRICA
SPAIN
SRI LANKA
SUDAN
SWEDEN
SWITZERLAND
SYRIAN ARAB REPUBLIC
TAJIKISTAN
THAILAND
TOGO
TONGA
TRINIDAD AND TOBAGO
TUNISIA
TÜRKİYE
TURKMENISTAN
UGANDA
UKRAINE
UNITED ARAB EMIRATES
UNITED KINGDOM OF
    GREAT BRITAIN AND
    NORTHERN IRELAND
UNITED REPUBLIC
    OF TANZANIA
UNITED STATES OF AMERICA
URUGUAY
UZBEKISTAN
VANUATU
VENEZUELA, BOLIVARIAN
    REPUBLIC OF
VIET NAM
YEMEN
ZAMBIA
ZIMBABWE

The Agency's Statute was approved on 23 October 1956 by the Conference on the Statute of the IAEA held at United Nations Headquarters, New York; it entered into force on 29 July 1957. The Headquarters of the Agency are situated in Vienna. Its principal objective is "to accelerate and enlarge the contribution of atomic energy to peace, health and prosperity throughout the world".

SAFETY REPORTS SERIES No. 117

REGULATORY CONTROL OF EXPOSURE DUE TO RADIONUCLIDES IN BUILDING MATERIALS AND CONSTRUCTION MATERIALS

INTERNATIONAL ATOMIC ENERGY AGENCY
VIENNA, 2023

COPYRIGHT NOTICE

All IAEA scientific and technical publications are protected by the terms of the Universal Copyright Convention as adopted in 1952 (Berne) and as revised in 1972 (Paris). The copyright has since been extended by the World Intellectual Property Organization (Geneva) to include electronic and virtual intellectual property. Permission to use whole or parts of texts contained in IAEA publications in printed or electronic form must be obtained and is usually subject to royalty agreements. Proposals for non-commercial reproductions and translations are welcomed and considered on a case-by-case basis. Enquiries should be addressed to the IAEA Publishing Section at:

Marketing and Sales Unit, Publishing Section
International Atomic Energy Agency
Vienna International Centre
PO Box 100
1400 Vienna, Austria
fax: +43 1 26007 22529
tel.: +43 1 2600 22417
email: sales.publications@iaea.org
www.iaea.org/publications

© IAEA, 2023

Printed by the IAEA in Austria
February 2023
STI/PUB/1992

IAEA Library Cataloguing in Publication Data

Names: International Atomic Energy Agency.
Title: Regulatory control of exposure due to radionuclides in building materials and
    construction materials / International Atomic Energy Agency.
Description: Vienna : International Atomic Energy Agency, 2023. | Series: IAEA
    safety reports series, ISSN 1020–6450 ; no. 117 | Includes bibliographical
    references.
Identifiers: IAEAL 23-01571 | ISBN 978–92–0–146522–1 (paperback : alk. paper) |
    ISBN 978–92–0–146622–8 (pdf) | ISBN 978–92–0–146722–5 (epub)
Subjects: LCSH: Radioisotopes — Safety measures. | Radioisotopes — Construction
    industry. | Radioactive substances.
Classification: UDC 539.163 | STI/PUB/1992

FOREWORD

Radiation exposure of the public may occur due to the presence of radionuclides in commodities such as building materials and construction materials. A systematic and graded approach is needed to control the exposure of members of the public to ionizing radiation emitted by these materials.

IAEA Safety Standards Series No. GSR Part 3, Radiation Protection and Safety of Radiation Sources: International Basic Safety Standards, establishes requirements for the regulatory control of radionuclides in a range of commodities, including construction materials. These requirements include the establishment of a legal and regulatory framework and the setting of reference levels for exposure resulting from radionuclides in commodities.

In November 2017, the Radiation Safety Standards Committee articulated the need to develop guidance on managing exposures arising from building and construction materials to support the application of the requirements of GSR Part 3 in conjunction with the recommendations provided in IAEA Safety Standards Series No. SSG-32, Protection of the Public Against Exposure Indoors due to Radon and Other Natural Sources of Radiation.

The roles and responsibilities of the parties involved, as well as practical mechanisms for demonstrating compliance with the reference level, are addressed in this Safety Report. The publication addresses both naturally occurring radioactive materials and artificial radionuclides that can unintentionally occur in building and construction materials.

The objective of this Safety Report is to provide practical guidance to governments, regulatory bodies, other relevant competent authorities, and the building and construction material industries on setting up arrangements for the regulatory control of building and construction materials that give rise to radiation exposures at all stages of the life cycle: raw material production, manufacturing, supply and end use. The responsibilities of the producers and suppliers of raw, recycled and reused materials for incorporation into building or construction materials are covered.

The IAEA officer responsible for this publication was O. German of the Division of Radiation, Transport and Waste Safety.

CONTENTS

1. INTRODUCTION

1.1. BACKGROUND

IAEA Safety Standards Series No. GSR Part 3, Radiation Protection and Safety of Radiation Sources: International Basic Safety Standards [1], establishes requirements for the protection of people against exposure to ionizing radiation (hereinafter termed radiation) and for the safety of radiation sources. The requirements of GSR Part 3 [1] are based on information on the detrimental effects attributed to radiation exposure provided by the United Nations Scientific Committee on the Effects of Atomic Radiation (UNSCEAR) [2] and the 2007 recommendations of the International Commission on Radiological Protection (ICRP) [3], and are intended to provide the basis for the regulation of sources of radiation.

Requirement 51 of GSR Part 3 [1] relates to protection of the public against exposure due to radionuclides in building and construction materials.[1] Paragraph 5.22 of GSR Part 3 [1] states:

> **"The regulatory body or other relevant authority shall establish specific reference levels for exposure due to radionuclides in commodities such as construction materials, food and feed, and in drinking water, each of which shall typically be expressed as, or be based on, an annual effective dose to the representative person that generally does not exceed a value of about 1 mSv."**

---

[1] The expression 'building and construction materials' is used in this Safety Report to make it searchable for construction experts as well as the radiological protection community, who are not necessarily familiar with each other's terminology. A definition of 'construction material' can also be derived from the European Construction Products Regulation [4], which defines a construction product as "any product or kit which is produced and placed on the market for incorporation in a permanent manner in construction works or parts thereof and the performance of which has an effect on the performance of the construction works with respect to the basic requirements for construction works".

The reference level[2] is used to optimize public exposure in existing exposure situations. The reference level of 1 mSv/a does not specify the origin of the radionuclides and is thus relevant for naturally occurring radioactive material (NORM)[3] as well as for artificial radionuclides, or a combination of both.

This Safety Report is intended to provide practical examples and detailed methods on how regulatory control of building and construction materials can be organized to allow for efficient methods of identification and restriction of exposure of the public due to NORM and artificial radionuclides.

The ICRP addressed exposure due to gamma emitting radionuclides of natural origin in building materials and in the ground in Ref. [6], which states that an "intervention exemption level of around 1 mSv is recommended for the individual annual dose expected from a dominant type of commodity, such as some building materials". Furthermore, the ICRP recommended that concerned national and, as appropriate, relevant international organizations should define the levels of exposure from commodities, in particular for specific building materials, at which certain actions need to be taken by national authorities.

1.2. OBJECTIVE

The objective of the Safety Report is to provide practical guidance to governments, regulatory bodies, other relevant competent authorities, and the building and construction industries on arrangements for the regulatory control and demonstration of compliance of building and construction materials that give rise to radiation exposures.

Guidance provided here, describing good practices, represents expert opinion but does not constitute recommendations made on the basis of a consensus of Member States.

----

[2] The concept of a 'reference level' is introduced by the ICRP in its 2007 recommendations [3] and implemented in the IAEA safety requirements [1] applicable to the regulatory control of commodities such as building and construction materials. Even though the application of the reference level under prevailing circumstances allows certain flexibility for national authorities in establishing higher or lower national regulatory values than those in GSR Part 3 [1], once the reference level has been defined in the national regulations or standards, compliance with it becomes mandatory.

[3] NORM is defined as radioactive material containing no significant amounts of radionuclides other than naturally occurring radionuclides [5]. NORM is a natural constituent of all stone-based building and construction materials due to their origin.

1.3. SCOPE

This Safety Report considers any building and construction materials that can be a source of radiation exposure to the public. Exposure resulting from such sources may include both external irradiation from gamma rays emitted by the building and construction materials and the inhalation of radon and thoron exhaled from such materials. In this Safety Report, however, only external irradiation from building and construction materials is considered. Recommendations on the regulatory control of radon in dwellings are provided in IAEA Safety Standards Series No. SSG-32, Protection of the Public Against Exposure Indoors due to Radon and Other Natural Sources of Radiation [7], and recommendations on protection against radon exposure in workplaces are in preparation [8].

This Safety Report covers protection against naturally occurring radionuclides, mainly from the uranium and thorium decay series and ^{40}K, as well as artificial radionuclides.

The Safety Report covers regulatory control of all the different possible steps in the lifetime of building and construction materials, such as approval of materials for the national market, raw material production, manufacturing, supply and end use. The responsibilities of the producers and suppliers of raw, recycled and reused materials for incorporation into building or construction materials are covered.

Verification of compliance (which is an integral part of the regulatory framework) for building and construction materials prior to their use, as well as in existing buildings and constructions, is also covered. Furthermore, methods for demonstration of compliance are discussed.

1.4. STRUCTURE

This Safety Report consists of four sections and four annexes. Section 1 sets out the objective and scope of the publication; Section 2 describes types of construction and building materials that could be of concern, as well as types of radiation that could be of concern; Section 3 describes the roles and responsibilities of governments and authorities and provides guidance to the regulatory bodies or other competent authorities in relation to regulatory control of building and construction materials; and Section 4 provides practical guidance to manufacturers, producers, suppliers and end users of building and construction materials on demonstration of compliance with the requirements.

The annexes provide examples of dose calculation models (Annex I), an overview of measurement instruments and methods (Annex II), information about building and construction materials entering the market (Annex III) and examples of regulations on radionuclides in building and construction materials (Annex IV).

2. IDENTIFICATION OF BUILDING AND CONSTRUCTION MATERIALS CONTAINING RADIONUCLIDES

Most building and construction materials contain some level of natural or artificial radionuclides. Their concentration depends on the origin of the materials, for example if they are of natural geological origin or if they have been exposed to contamination, either from past practices or in the course of authorized activities.

It is the responsibility of the government, or government appointed authorities, to identify which types of building and construction materials, or their constituents, need to be regulated for radiation protection purposes. This information needs to be made available, for example, in regulatory databases, in dedicated internet resources or in publications.

This section provides practical guidance on the identification of building and construction materials that may be of concern and their possible origin, as well as on the provision of information.

2.1. IDENTIFICATION OF BUILDING AND CONSTRUCTION MATERIALS CONTAINING RADIONUCLIDES

Natural radioactivity is ubiquitous: it occurs in different concentrations in the Earth's crust and thus also in minerals and residues from industrial processes that are commonly and widely used as either building and construction materials or as constituents or additives to building and construction materials. Consequently, building and construction materials are a source of radiation exposure to the population. They account for between one third and two thirds of the overall radiation exposure of the population in some States [9].

Although radiation exposure from natural radioactivity in building and construction materials is generally unavoidable, the radionuclide content of such materials and the resulting doses may vary widely, depending on the geological origin of any natural materials used and the presence of residues from industrial processes employed in the production of building and construction materials. If contributions from radon and thoron are excluded, the worldwide average indoor radiation exposure (population weighted) is approximately 0.4 mSv/a.

The vast majority of building and construction materials do not lead to exposures above the reference level of 1 mSv/a. However, in some cases, the concentrations of radionuclides in building and construction materials do result in levels of exposure that are generally considered to be of concern from a radiation protection point of view (i.e. above 1 mSv/a). In the UNSCEAR review of doses and sources of radiation [10], effective doses of the order of 10 mSv/a or even higher (external irradiation only) were reported in some States.

2.2. NATIONAL SURVEY OF NATURAL RADIOACTIVITY IN BUILDING AND CONSTRUCTION MATERIALS

The government of a State may initiate representative surveys of radiation exposures resulting from building and construction materials used in the State and define the principal materials that have the potential to give rise to exposures above the reference level.

The surveys need to cover raw, reused and recycled materials used for the production of construction and building materials. This includes any materials produced within the State, as well as materials imported from other States.

On the basis of such overviews, the government or a designated competent authority can compile a list of materials it wants to regulate. Examples of such materials could be as follows:

(a) Materials of natural origin: alum shale, building and construction materials or additives to them of natural igneous origin, such as granitoids (e.g. granites, syenite, orthogneiss), porphyries, tuff, pozzolana (pozzolanic ash) and lava.
(b) Materials incorporating residues from industries processing NORM: fly ash, phosphogypsum, phosphorus slag, tin slag, copper slag and red mud (residue from aluminium production).
(c) Residues from steel production.
(d) Conditionally cleared materials to be reused in the construction industry.
(e) Metal components of building and construction materials that are produced from recycled scrap.

This set of examples is neither exhaustive nor obligatory, and needs to be extended to include any other materials of concern identified in the individual State.

The survey also needs to include a retrospective overview of whether any of the identified materials have been widely used in the State and could be a cause of public exposure in existing buildings.

It is good practice to periodically review the established list of materials because changes of circumstances and materials may occur. Consideration needs to be given to traditional materials that may be commonly used, but only in certain regions of the State.

Special consideration is also needed for the list of building and construction materials to be regulated in States that mainly import such materials, as products of the same type might have significantly varying characteristics, depending on their origin.

Surveys may be based on screening tools provided by the regulatory body, for instance, usually using conservative models and parameters. Further details are provided in Section 3.3 and Annexes I and II.

2.3. ARTIFICIAL RADIONUCLIDES IN BUILDING AND CONSTRUCTION MATERIALS

Some of the most common sources of artificial radionuclides in building and construction materials are radioactive sources that were melted during steel production. It is a common practice to monitor scrap metal at manufacturing facilities, as well as imported metal [11]. This may be especially important for countries that include areas prone to earthquakes, where a lot of steel and metal is used in the construction of buildings. In general, building and construction materials, or their constituents, only need to be checked for artificial radioactivity where it is suspected that artificial radionuclides might be the cause of undue exposure from external irradiation.

Several incidents of significant amounts of artificial radionuclides either being discovered in raw building and construction materials or being found in existing buildings have been documented worldwide. In one incident in Taiwan, China, a ^{60}Co source was melted during rebar manufacturing. Subsequently, 181 buildings were constructed using contaminated rebar. The contamination was only discovered ten years later and led to an estimated average excess cumulative dose of approximately 50 mSv [12].

A different type of incident occurred in Ukraine, in which a capsule containing radioactive ^{137}Cs was found inside the concrete wall of an apartment building. The source, originally a part of a radiation level gauge, had been lost in a quarry. Gravel from the quarry was used to produce concrete for the residential building, where the source was located in the wall of one of the bedrooms. Two families lived in this apartment for more than nine years, and by the time the capsule was discovered the overexposure had produced drastic results [13].

The authorities need to consider all relevant circumstances before deciding on the further use of contaminated materials. If the decision is to use the contaminated materials (e.g. owing to lack of supply), the reference level applies to the sum of the contribution from natural and artificial radioactivity. If necessary, the use of the materials for construction purposes could be restricted, which leads to the establishment and implementation of a supervision programme by the competent authority.

2.4. REUSE OF CLEARED MATERIAL IN BUILDING AND CONSTRUCTION MATERIALS

Radiation exposure of the public from building and construction materials is mainly the result of enhanced levels of natural radioactivity. However, artificial radionuclides can occur in building and construction materials because of past practices, accidental contamination or the reuse of cleared or exempted material [14].

Paragraph I.11 of GSR Part 3 states:

"Material may be cleared without further consideration...provided that in reasonably foreseeable circumstances the effective dose expected to be incurred by any individual owing to the cleared material is of the order of 10 µSv or less in a year. To take into account low probability scenarios, a different criterion can be used, namely that the effective dose expected to be incurred by any individual for such low probability scenarios does not exceed 1 mSv in a year."

For radionuclides of natural origin, para. I.12(c) of GSR Part 3 [1] states:

"For radionuclides of natural origin in residues that might be recycled into construction materials...the activity concentration in the residues does not exceed specific values derived so as to meet a dose criterion of the order of 1 mSv in a year, which is commensurate with typical doses due to natural background levels of radiation."

The general clearance levels presented in table I.3 of GSR Part 3 [1] for radionuclides in the uranium decay chain or the thorium decay chain (1 Bq/g) and for ^{40}K (10 Bq/g) are generally too high to ensure that building and construction materials comply with the 1 mSv/a dose criterion. As such, additional restrictions on the reuse of such materials in building and construction materials may be necessary.

Higher levels of artificial radioactivity in building and construction materials may arise from unplanned events. In such cases, specifically derived clearance levels, which allow for higher annual exposures, might be considered. However, if these materials are to be used for building and construction purposes, the same reference level of 1 mSv/a applies. There is a need to establish a special procedure for verification of the intended use of specifically cleared materials [15].

The authorities and the building and construction industry need to consider the public interest in these types of materials.

The control of materials that are cleared is normally applied at the time and location when clearance is granted. These materials may then be used freely without restrictions [1, 14]. If conditional clearance is allowed by the national legal framework, special care needs to be taken that these materials are only used for their intended purpose and, for example, are not accidentally used for the construction of buildings [15]. The building and construction industry needs to be aware, when purchasing such materials for reuse and recycling, that artificial radionuclides could be subject to verification of compliance. In particular, the regulatory body may wish to check the use of conditionally cleared materials.

2.5. RADIONUCLIDES IN MATERIALS OF EXISTING BUILDINGS

Buildings constructed before the requirements of GSR Part 3 [1] were established may contain materials with high concentrations of NORM or artificial radionuclides that could lead to exposures above the 1 mSv/a reference level. For example, in the Czech Republic, Romania and Sweden, materials with elevated levels of radium were used for the construction of dwellings during the 1970s [16, 17]. Also, unauthorized use of mine tailings occurred in the areas around uranium mines, where civilians used the tailings as backfill in the foundations of their homes.

The same level of protection for the population, such as the same reference level, applies to all buildings, irrespective of the year of construction. If there is a reasonable suspicion that materials with elevated levels of NORM or artificial radionuclides were used in the construction of buildings, special surveys may need to be undertaken to identify such buildings, and remediation actions, if justified, need to be taken. Further guidance on the remediation of buildings with high gamma exposure and high radon concentrations is provided in Ref. [18].

For privately owned dwellings, the identification of buildings with high levels of exposure from gamma radiation, as well as high radon activity concentrations, may occur at the time of a real estate transaction, when either a potential buyer requests a measurement certificate or a building's owner provides it voluntarily. This practice could also be made a mandatory regulatory requirement to ensure radiation safety and consumer rights protection.

2.6. PROVIDING INFORMATION ON RADIONUCLIDES IN BUILDING AND CONSTRUCTION MATERIALS

Paragraph 5.5 of GSR Part 3 [1] states:

"The regulatory body or other relevant authority shall implement the protection strategy, including:

....

(b) Ensuring that information is available to individuals subject to exposure on potential health risks and on the means available for reducing their exposures and the associated risks."

It is the responsibility of the government to make available information on building and construction materials that are of concern from a radiation protection point of view. This information can be identified in accordance with the process described in this section and made available to the public and other interested parties.

For artificial radionuclides, a regulatory framework for conditional clearance may need to be established, as appropriate, and sellers and buyers of such materials need to be aware of the conditions for reusing conditionally cleared materials. These conditions will be based on specific exposure scenarios to ensure that the dose to the public will not exceed the reference level [14, 15].

For NORM, the derived activity concentrations for building and construction materials (see para. I.12(c) of GSR Part 3 [1]) will generally be more stringent than general clearance levels (i.e. 1 Bq/g for ^{232}Th and ^{238}U decay series and 10 Bq/g for ^{40}K; see table I.3 of GSR Part 3 [1]). If a material has been cleared on the basis of these general clearance levels, it does not automatically mean that the material will be allowed for use in construction, especially in the construction of dwellings.

In many States, labelling and certification of imported materials is mandatory in order to provide information on the origin of the material and its radiological and other characteristics. However, this does not guarantee that material approved in one State will also be accepted by another State because they may take different regulatory approaches, and the models used for dose calculations and their underlying assumptions on parameters may be different.

States may strive for international harmonization of the approach to building and construction materials. For example, a harmonized approach in Nordic countries is described in Ref. [17]. Governments and competent authorities need transparency concerning the radiological assessment of building and construction materials or, ideally, the representative measurement results (original data on the radionuclide content from the exporting State) in order to allow the importing State to check whether the relevant regulatory requirements are met.

3. RESPONSIBILITIES OF THE GOVERNMENT AND AUTHORITIES

Requirements on the responsibility of the government in the field of radiation protection and radiation safety are established in IAEA Safety Standards Series No. GSR Part 1, Governmental, Legal and Regulatory Framework for Safety [19]. Requirement 3 of GSR Part 1 [19] states:

"The government, through the legal system, shall establish and maintain a regulatory body, and confer on it the legal authority and provide it with the competence and the resources necessary to fulfil its statutory obligation for the regulatory control of facilities and activities."

Requirement 4 of GSR Part 1 [19] states that "The government shall ensure that the regulatory body is effectively independent in its safety related decision making".

In this Safety Report, it is assumed that a legal and regulatory framework meeting the requirements of GSR Part 1 [19] already exists, and that the regulatory framework for the protection of the public against indoor exposures from building and construction materials can essentially be based on that existing system. This includes an effectively independent regulatory body with specified responsibilities and functions or another competent authority with responsibilities for these types of material. It is also assumed that a regulatory system for the control and approval of building and construction materials is already in place.

Therefore, this section only considers specific requirements and challenges from exposure due to radionuclides in building and construction materials. It also outlines the regulations and requirements that need to be enforced by the regulatory body or other competent authority in order to ensure the safety of building and construction materials. The competent authority could be a public health authority, a building and construction safety authority, a radiation protection authority or another authority that has been assigned responsibility for the protection of the population against radiation exposure arising from building and construction materials.

3.1. LEGAL FRAMEWORK FOR THE CONTROL OF RADIONUCLIDES IN BUILDING AND CONSTRUCTION MATERIALS

In accordance with Requirement 2 of GSR Part 3 [1], the government is responsible for establishing the legal framework for protection and safety and for establishing a regulatory body with responsibilities for the development of regulations and their enforcement.

As stated in Section 1.1, Requirement 51 of GSR Part 3 [1] requires that the regulatory body or other relevant authority establish reference levels for building and construction materials. It is normally not considered adequate to establish a reference level of effective dose to members of the public from external irradiation from building and construction materials that exceeds 1 mSv/a. This excludes exposures due to radon, for which there will be a separate reference level expressed in terms of the annual average activity concentration of radon (see Requirement 50, para. 5.20(a) of GSR Part 3 [1]).

If a national or regional survey (see Section 2.2) shows that the building industry of the State (or a region within a State) relies heavily on construction materials that regularly contain enhanced levels of natural radioactivity and lead to effective doses to the representative person above 1 mSv/a, reference levels higher than 1 mSv/a might be considered, taking into account health protection, social, economic, environmental and cultural aspects. It is also a responsibility of the government to communicate such a decision to its citizens and to relevant material producers.

Building and construction materials are also subject to other, industry specific, regulations and standards. These include comprehensive measures to regulate and standardize the physical, chemical and biological properties of construction materials [4, 20]. Establishing additional legal frameworks for radiation protection may therefore involve consultations with the relevant competent authorities in order to effectively integrate radiation protection goals by merging the existing regulatory and implementation systems in a functional and effective way and, in so doing, making effective use of the existing regulatory frameworks.

In the European Union (EU), the Construction Products Regulation [4] forms the basis for the regulation of different properties of building materials that are relevant to health and safety. This includes protection against ionizing radiation. The Construction Products Regulation is reflected in national regulations, standards and building codes of the member states of the EU.

In Canada, standards for testing and requirements on certain characteristics of building and construction materials are in place, including a specified frequency of testing [21, 22]. These standards are then referred to in the Canadian building code [23].

Care is needed to clearly define the fields and boundaries of responsibility of the different government ministries, authorities, producers of building materials, building companies, architects and others in order to avoid conflicts of interest or gaps in regulation. For example, a regulatory body may be responsible for the establishment of regulations, whereas the construction safety authority may be responsible for the verification of compliance and enforcement of these regulations with regard to building and construction materials. Alternatively, the construction safety authority may also be assigned sole responsibility for both tasks.

The Radiation Protection Act in Germany, together with the related Radiation Protection Ordinance [24, 25], sets the requirements for building and construction materials, including a reference level of 1 mSv/a. The verification of compliance is, however, performed by the Institute of Building Engineering (Deutsches Institut für Bautechnik), which is generally responsible for the approval of building and construction materials in Germany.

It is the responsibility of the government to provide the financial, technical and human resources necessary to reliably and effectively implement a system to control radiation risks (see Requirement 3 and paras 2.3 and 2.5 of GSR Part 1 [19]), including for building and construction materials concerning radionuclides. For instance, the implementation of the requirements on building materials needs specific human and technical resources, such as qualified experts for reliable dose assessment and independent laboratories that are able to measure the radioactivity content of building materials.

3.2. GRADED APPROACH TO THE CONTROL OF RADIONUCLIDES IN BUILDING AND CONSTRUCTION MATERIALS

Controlling every kind of building and construction material for radiation protection purposes would be neither feasible nor necessary. It is therefore advisable to use a graded approach, where only building and construction materials that could give rise to exposures above the established national reference level are regulated.

The concentration and type of radionuclides of natural origin in minerals used in the production of building and construction materials depend on the geology of the place of origin and on the individual rock body characteristics. The concentrations of radionuclides of natural origin in residues from industrial processes (e.g. phosphogypsum, fly ash, slag from steel plants) reused in the building and construction industry also depend on the origin of such residues.

Thus, it may be useful to preselect materials and industrial residues that could be of relevance for radiation protection purposes as part of a graded approach. As an example of such an approach, the European Commission created a list of materials likely to become subject to regulatory control in annex XIII to Ref. [26].

In some States, such as Brazil, less comprehensive approaches are used to establish the scope of regulation of building and construction materials, including the reuse of specific materials, such as phosphogypsum, in agriculture and the cement industry. Reference [27] provides an example of the establishment of exemption levels for the use of phosphogypsum in agriculture or in the cement industry in Brazil. A maximum activity concentration of 1 Bq/g is set for each radionuclide ^{226}Ra and ^{228}Ra, based on the recommendations provided in IAEA Safety Standards Series No. RS-G-1.7, Application of the Concepts of Exclusion, Exemption and Clearance [14]. Guidance for the construction industry is provided by the regulatory body, including on the administrative and technical procedures relating to Refs [27, 28].

Another example of a graded approach to the regulatory control of building and construction materials can be found in Russian regulations, where a classification of building and construction materials into four classes is used [29]. This classification is carried out using the results of activity concentration measurements of gamma emitting NORM radionuclides. The four classes of building and construction materials are used to define restrictions on the use of the materials, such as in residential, public and industrial buildings and facilities, and in road construction inside or outside town borders. The fourth category of the materials can be used only with the permission of the Federal authorities, subject to a specific dose assessment.

3.3. REGULATORY FRAMEWORK FOR THE CONTROL OF RADIONUCLIDES IN BUILDING AND CONSTRUCTION MATERIALS

It is important to establish a regulatory framework that addresses all causes of exposure to the public due to building and construction materials and that identifies the parties responsible for compliance (e.g. manufacturers, builders). The following subsections address who the responsible parties may be, how they may demonstrate that their building and construction materials are in compliance with the regulations and what happens if they do not comply. Furthermore, the concept of a screening tool is introduced as a possible practical instrument to make the demonstration of compliance more efficient.

Establishing regulatory requirements for building and construction materials, including a reference level not exceeding 1 mSv/a, is important for both existing buildings and newly constructed ones. The construction industry's main responsibility is for the current uses of building and construction materials, and compliance can be expected from the time that the regulations are enacted. For existing buildings, the reference level may be used to support the protection of the public and to enforce corrective actions in cases where the reference level is exceeded.

3.3.1. Regulations and guidance on building and construction materials containing radionuclides

Once a reference level has been established by the regulatory body or other authority assigned by the government, regulations need to define the parties responsible to ensure that building and construction materials are in compliance. Producers and manufacturers of construction materials, importers, traders and construction companies could be considered the responsible parties at different stages of the life cycle of such materials and could therefore be responsible for demonstrating compliance with regulations.

The regulatory body, or other authority, may also issue guidance documents to support the implementation of the requirements. When introducing new regulatory requirements and guidance on radiation protection for building and construction materials, it is important to verify that the responsibility of compliance is assigned to both manufacturers and end users, as well as the importers of such materials. For example, if regulations on building and construction materials are only included in the relevant building regulations, only the building and construction companies will be held responsible for complying with the reference level. It is, however, good practice to also give responsibilities to the producers and manufacturers of construction materials to ensure that only building and construction materials that are in compliance with the reference level are sold and used. This may be achieved by placing legal product requirements on building and construction materials. As such, producers and manufacturers would only be allowed to sell construction materials that had been shown to comply with the reference level, and builders and construction companies would only be allowed to use building materials that were in compliance. This combination of a 'bottom-up' (producers) and 'top-down' (builders) approach may help to ensure compliance within a fast paced building market.

The regulations and guidance that control radioactivity in building and construction materials may, for example, be issued in the areas of radiation protection or public health protection, or as building codes. They are established or adopted by the regulatory body (or other competent authorities) and can address the reference level in one of the following ways:

(a) By establishing derived activity concentrations for radionuclides;
(b) By using an activity index[4];
(c) By establishing derived reference levels in terms of gamma dose rate.

Establishing derived reference levels in terms of the gamma dose rate in a building may serve two purposes:

(a) To check existing buildings when there is a reasonable suspicion that inappropriate materials were used in their construction;
(b) To verify that no radioactive materials or sources were accidentally incorporated into buildings during construction.

The regulations also need to describe requirements for measurement quality, record keeping of the measurement results, and the form and frequency of reporting.

Materials originating from facilities using nuclear material or other radioactive material are subject to regulatory requirements for the clearance of radioactive materials. These requirements are usually enforced by the regulatory body for radiation safety.

3.3.2. Demonstration of compliance with requirements for building and construction materials containing radionuclides

To demonstrate compliance with regulatory requirements for building and construction materials containing radionuclides, a producer, importer, trader or end user of the materials needs to perform radiological measurements. This can be done either within the organization or by an independent measurement service provider. In either case, measurements need to comply with quality assurance requirements and use an approved methodology, as described in Sections 3.4 and 3.5.

---

[4] An 'activity index' is an index that is derived from the activity concentration of the radionuclides that may be present in building and construction materials. This is intended as a screening tool to estimate whether a building or construction material complies with the dose reference level. A 'screening tool' is a simplified means to define the appropriateness of building and construction materials using activity concentration measurements. It is based on a conservative model, thus ensuring compliance with the reference level.

The samples used for these measurements need to be representative of the construction and building material in question. Because of the inhomogeneous nature of some raw materials, special sampling procedures may have to be considered (e.g. for sampling bulk materials, or for when the supplier of a raw material is changed). Several industry standards exist that define the frequency of testing of such materials for other characteristics and properties [30–33].

In some States, quarries mining raw materials intended for the production of building and construction materials are categorized in order to determine whether the material is appropriate for residential or public buildings. However, quarries might not be evaluated as a whole because of the inhomogeneous characteristics of the rock body. Separate testing of produced blocks or batches in development is needed to demonstrate compliance.

A more complex approach whereby buildings are categorized according to their occupancy time is applied in some States; in China, for example, buildings are classified into one of two categories based on their type and corresponding occupancy time (see Annex IV).

If there is a need to demonstrate compliance with the reference level for existing buildings, gamma dose rate measurements are the appropriate method to use (see Annexes I and II).

The regulations may also include requirements for the format and timing of records that need to be kept in order to demonstrate compliance. These requirements could be harmonized with the other requirements for building and construction materials, consumer rights and other relevant regulations.

3.3.3. Establishment of a screening tool for building and construction materials containing radionuclides

The reference level providing the basis for national regulations is based on the annual effective dose value of 1 mSv. The calculation of the annual effective dose may be complex and can only be completed by radiation protection experts (see appendix VI in Ref. [17]). It is therefore a very common practice for a screening tool to be incorporated into the regulations or guidance in order to provide a simpler means of demonstrating compliance with the reference level.

The screening tool may be a simple calculation of an activity index, based on measured activity concentrations of ^{40}K, ^{226}Ra and ^{232}Th. One example of such a formula is the following calculation of the activity index (denoted AI in equation (1) below), as recommended by the radiation protection authorities of Nordic countries in 2000 [17]:

$$AI = \frac{C_{Th-232}}{200} + \frac{C_{Ra-226}}{300} + \frac{C_{K-40}}{3000} \qquad (1)$$

where C_{Th-232}, C_{Ra-226} and C_{K-40} are the activity concentrations of ^{232}Th, ^{226}Ra and ^{40}K, respectively, expressed in Bq/kg.

If the result of this calculation is below a threshold set in the national regulations or guidance, the use of the construction material is expected to comply with the reference level of 1 mSv/a.

The assumptions and conditions on which a screening tool is based need to be reasonably conservative and representative of the State in which they are used. For composite construction materials, the screening tool is to be applied to the final mixture of the building and construction material, not to its individual constituents. In cases where the final product may not be easily assessed, it may be sufficient to apply an appropriate partitioning factor to the results of the screening of all its individual constituents.

If raw materials are used in construction without further mixing or processing, the screening tool applies to them in the same manner.

Where construction materials are not used as a bulk material but, for example, as a superficial or decorative material, such as tiles, gypsum boards or granite decorations, a different threshold for the activity index may be applicable. For example, in China, the Czech Republic and Finland, superficial materials are addressed separately from bulk materials. If the construction material exceeds this modified threshold, it may still be used in a restricted manner, as described in national regulations.

More examples of national regulations are provided in Annex IV.

3.3.4. Non-compliance of building and construction materials containing radionuclides

If non-compliance is discovered, the regulatory body or competent authority will need to be notified by the responsible parties within the time frame established in national regulations or guidance. The regulatory framework therefore needs to include details of the notification responsibility, the information to be submitted, the time frames for notification and the identification of which authority is to be notified.

3.3.4.1. When to notify the competent authority

Regulations or guidance need to specify the conditions of notification of competent authorities, including which authority is to be notified. In general, notification may be appropriate in the following cases:

— A new building or construction material is put on the market and is assessed as giving rise to radiation exposures above 1 mSv/a (notification by the manufacturer, importer or trader — to be defined in the regulations);
— An existing building or construction material is assessed as giving rise to radiation exposures above 1 mSv/a because of changes in the constituents (notification by the manufacturer or trader);
— An existing building is assessed as giving rise to radiation exposures above 1 mSv/a (notification by the building owner, which could be a construction company or developer — to be defined in the regulations).

3.3.4.2. Information to be submitted in the notification

For building and construction materials, information to be submitted in the notification may include the following:

— Place of origin of the building or construction material.
— Supplier, trader, importer or manufacturer name and address.
— Activity concentration of all relevant radionuclides determined by a qualified in-house laboratory or an external measurement service provider.
— Cause of non-compliance:
  • Comparison of the measured activity concentration with the values established in the regulations or guidance;
  • A dose assessment indicates that the established reference level is exceeded;
  • Results of the analysis of the cause of non-compliance.
— Intended use of the construction and building materials.
— Justification of the use of this construction or building material.

For existing buildings, such information may include the following:

— Name and address of construction company, developer or building owner.
— Address of the building.
— Current and intended use of the building.
— Records of the dose rate measurement results in different rooms or locations within the building.

— Cause of non-compliance:
  - The measured dose rates or activity concentrations exceed the values established in the regulations or guidance;
  - A dose assessment indicates that the established reference level is exceeded;
  - Results of the analysis of the cause of non-compliance.
— Proposed corrective measures and follow up measurements.

The notified authority will need to follow up on cases of non-compliance in the use of the building and construction materials and, where appropriate, in existing buildings.

If a building or construction material does not comply with the regulatory requirements established for the intended type of building, it may normally not be used. Any provisions by the authorities to allow it to be used will need to be justified.

If it is discovered that a newly constructed building does not comply with the regulatory requirements, appropriate corrective actions need to be implemented and compliance demonstrated before it is used for its intended purpose.

If it is discovered that an existing building does not comply with the regulatory requirements, decisions on the appropriate next steps need to be taken on a case by case basis [18]. The responsibility for corrective actions in such buildings needs to be defined in the national legal system or regulations.

The starting point is usually to identify the party responsible for the non-compliance and enforce appropriate corrective actions. In cases where such a party does not exist or cannot be identified, or where the building was commissioned before the current regulations were put in place, the owner of a building is usually responsible for the necessary corrective actions. However, other national mechanisms may be used to enforce corrective actions, if available and justified (e.g. remediation funds).

3.4. REGULATORY REQUIREMENTS FOR QUALIFIED IN-HOUSE LABORATORIES AND EXTERNAL MEASUREMENT SERVICE PROVIDERS

Measurements of radioactivity in building and construction materials and of gamma dose rates in existing buildings need to be performed by a laboratory that fulfils the requirements set by the regulatory body or other competent authority. Such requirements may include accreditation by a national body or approval by the relevant authorities. The regulations or guidance need to include the relevant requirements.

Large producers of building and construction materials that possess their own qualified laboratories may be obliged to send a certain percentage of samples to independent measurement service providers to confirm or validate the results of their measurements.

An appropriate system of quality assurance needs to be in place in order to ensure that all measurement results are accurate and traceable to national or international standards. This includes the following:

— Documentation (e.g. of sample arrival, condition of sample, weight, start of measurement, measurement time, detector system that was used, uncertainty analysis);
— Traceability of measurement results to national/international standard sources (calibration of measurement equipment);
— Calibration of scales and measurement equipment, regular performance tests;
— Demonstration that measures have been taken to avoid cross-contamination among samples during sample preparation (standardized procedure descriptions).

Recommendations for providers of such technical services are provided in section 8 of IAEA Safety Standards Series No. GSG-7, Occupational Radiation Protection [34].

3.5. MEASUREMENT METHODS FOR BUILDING AND CONSTRUCTION MATERIALS CONTAINING RADIONUCLIDES

States need to establish (or provide guidance on) procedures and methods of dose assessment to ensure consistency. Depending on whether the exposure is due to an existing building or due to building and construction materials, different measurement methods are needed. Depending on the activity concentrations of the radionuclides within the building and construction materials, the annual effective dose to the representative person may be estimated using a model that takes into account parameters such as the thickness and density of the material and the occupancy time.

For existing buildings, measurements of the gamma dose rate can be used to estimate the annual effective doses to occupants. It is advisable that the same occupancy factor for different types of buildings be used for dose estimation, as the use of buildings may change (e.g. a public building with low occupancy time may in the future be converted into a residential building with a high occupancy).

3.5.1. Determination of the activity concentration of radionuclides in building and construction materials

The accurate determination of the activity concentration of radionuclides in building and construction materials is necessary for dose assessments and for the application of screening tools. Different methods are available to determine the activity concentration of radionuclides in building and construction materials. The most commonly applied method is gamma spectrometry (see Annex II). The regulatory body may want to provide guidance on the lower limit of detection and/or the uncertainty of the measurement results.

All samples submitted for measurement need to be representative of the building or construction material in question. High quality measurement results are of little use if the samples do not reflect the actual radioactivity content inside the materials. Homogenization as part of the sample preparation could be necessary only if it is representative of the material in its intended form of use.

Usually, measurement of the activity concentration of the naturally occurring radionuclides ^{40}K, ^{226}Ra and ^{232}Th will be the main focus. Although the determination of ^{40}K is very straightforward (a single gamma emission at 1460 keV), this is not the case for ^{226}Ra and ^{232}Th, as the intensity of their gamma emissions is low, and interference with other radionuclides can make their direct determination more difficult (e.g. ^{235}U may interfere with the analysis of ^{226}Ra gamma emission at 186 keV). In addition, equipment with a detector volume large enough to ensure a satisfactory limit of detection within a reasonable counting time is needed.

Gamma spectrometry equipment needs to have a traceable calibration (see Ref. [35]).

Alternatively, ^{226}Ra and ^{232}Th can be determined from the gamma emissions of their daughter products. If secular equilibrium between ^{232}Th and its daughter products is likely, it may be determined through the measurement of ^{228}Ac. Radium-226 may be determined through the measurement of ^{214}Pb. However, ^{226}Ra decays into ^{214}Pb via ^{222}Rn, which may escape from the sample before it decays into its daughter products. For this reason, special care has to be taken to ensure secular equilibrium between ^{226}Ra and its daughters. This measurement method requires more sample preparation (e.g. sealing sample containers to prevent radon escape) and a processing time of approximately 28 days to establish the equilibrium before measurement.

The measurement methods summarized in Annex II provide details of the two different measurement approaches.

In some cases, the activity concentration of artificial radionuclides such as ^{137}Cs or ^{60}Co may need to be measured (e.g. because of accidents with radioactive sources or when materials are recycled). Caesium-137 and ^{60}Co are readily measured by gamma spectrometry equipment available on the market today, allowing for high quality results of such measurements. Also, a reliable quality assurance system and calibration services exist for measurements of radiocaesium and cobalt.

3.5.2. Determination of gamma dose rates in buildings containing radionuclides

In an existing building, direct measurement of the gamma dose rate may be used to demonstrate compliance with the dose reference levels for building and construction materials. Measures need to be taken to ensure that dose rate measurements from different buildings can be compared, taking into account contributions from surrounding materials and the position of the dose rate measurement.

Two approaches can be used:

(a) Direct gamma dose rate measurement and direct calculation of annual effective dose. This is a simple and efficient way of demonstrating compliance, especially if a derived reference level in terms of the gamma dose rate is set in regulations or guidance.
(b) Measurement of the activity concentration of naturally occurring radionuclides to calculate the dose rate due to NORM. This method can be used to demonstrate compliance in cases where the derived reference level set in the regulations is expressed in terms of the dose rate from NORM.

Practical information on the measurement of the dose rate inside existing buildings can be found in Annex II.

4. DEMONSTRATION OF COMPLIANCE WITH REGULATORY REQUIREMENTS FOR RADIONUCLIDES IN BUILDING AND CONSTRUCTION MATERIALS

This section outlines the responsibilities of manufacturers, suppliers and end users of building and construction materials to ensure that these materials meet the relevant regulatory requirements for the protection of the public against exposure due to radionuclides in these materials.

It is good practice for producers and traders of building and construction materials to distribute only building and construction materials that are in compliance with the requirements defined by the competent authorities. In many States, an activity index is used as a screening tool to ensure that exposures do not exceed reference levels. Individual States may also specify derived reference levels, in terms of the activity concentration of radionuclides in building and construction materials, on the basis of where the materials are to be used (e.g. in residential buildings, industrial buildings, road construction or landfill) and how the materials are to be used (e.g. as bulk materials, in superficial or decorative materials, or as a percentage of hollow parts like windows or doors).

Many States specify derived reference levels for building and construction materials in terms of the activity concentrations of naturally occurring radionuclides or in terms of the gamma dose rate in the centres of rooms in existing buildings.

Manufacturers, producers, suppliers and end users of building and construction materials need to comply with the relevant requirements established in the national legal or regulatory framework, as described in Section 3.

4.1. MANUFACTURERS OF BUILDING AND CONSTRUCTION MATERIALS

Manufacturers and producers of building and construction materials containing radionuclides potentially include providers of lumber, concrete, minerals, clays, natural and engineered stones, bricks, flooring products and construction metals. As in many other industries, manufacturers of building and construction materials provide their company name, address and contact information for purposes of traceability when supplying these materials, and the materials are normally tested and evaluated by qualified laboratories for purposes other than radiation protection (chemical and physical properties). For radiation protection purposes, it is useful for building and construction materials to be clearly labelled or documented, specifying the name, origin and constituents of the product, as well as their intended use.

Manufacturers are responsible for the control of raw materials prior to their use in building and construction materials and need to take into consideration factors such as the type of material and where it is from. Geological considerations will often be a good general guide to identifying areas in which enhanced activity concentrations of radionuclides of natural origin are likely to be found. Manufacturers may ask the relevant competent authority for more information.

Caution needs to be exercised by manufacturers and producers to avoid using tailings from uranium mining sites in building and construction materials.

In fact, special care needs to be taken by producers when they are reusing residues which come from any NORM industry or which might contain artificial radionuclides. For example, in Australia, although the majority of the material generated from mining and processing is disposed of, some materials may be used for road construction or as a constituent of building and construction materials. In such cases, the materials can be used only after approval from the relevant competent authority [14, 15].

Manufacturers and producers need to ensure that representative samples (see Section 3.5.1) are taken to determine the activity concentration of radionuclides in building and construction materials. For example, the minimum rate of sampling for assessing the compressive strength of concrete recommended in Ref. [20] is one sample for every 400 m³ (or one sample every production week). These samples can also be used for radioactivity measurements. The sampling strategy and frequency can also be influenced by other aspects, such as the origin of the materials and the variation in activity concentrations.

If the activity concentrations of radionuclides are measured in the raw materials used to produce a building or construction material, the producer can use these results to estimate the activity concentrations of radionuclides in the end product on the basis of the mixing ratios of raw materials and the mass changes during production processes, etc. It is the producers' responsibility to verify that the estimated activity concentrations and the concentration index of their products are below the permitted values defined by the relevant competent authorities.

Producers of building and construction materials need to have the activity concentration in their products measured by a qualified laboratory. This could either be their own laboratory or an independent measurement service provider. The measurement methods and dose estimation methods used need to comply with any regulations or guidance provided by the competent authorities (see Sections 3.4 and 3.5). Compliance with an activity index (see Section 3.3.1) can also be calculated using the measured activity concentrations. The results of the measurements need to be documented and records kept for as long as required by the regulations. The records need to be made available to competent authorities and buyers of the material when requested.

If artificial radionuclides are suspected to be present in building and construction materials, the producers need to report this to the relevant authorities and consult qualified laboratories to further identify the radionuclides and investigate the radioactivity in the material. If materials have been conditionally cleared for reuse or recycling as a building or construction material, the producers need to include the intended purpose of the materials on the labelling or in the technical specifications of the material.

4.2. SUPPLIERS OF BUILDING AND CONSTRUCTION MATERIALS

Suppliers of building and construction materials, including traders, distributors and importers, are encouraged to ensure the following:

— That only labelled materials with technical specifications are made available;
— That measurement of activity concentrations of radionuclides has been made by qualified laboratories (laboratory testing may be provided by the manufacturer, producer or importer of the materials or by an external service provider);
— That building and construction materials comply with any applicable activity index defined in national regulations or guidance.

Documentation of demonstrated compliance can be included as one of the essential requirements on material properties in any trade agreements.

When suppliers purchase or sell building and construction materials internationally, they become importers or exporters and need to also ensure that the materials meet the relevant requirements and regulations in other States.

Importers and exporters need to note that, even if the activity concentrations of the radionuclides in a building or construction material are below general clearance levels, such materials might still cause the reference levels defined in individual States to be exceeded (see also Section 2.4).

Suppliers of building and construction materials are encouraged to share the results of radiological measurements and dose assessments with their customers, including other traders or end users, on request.

For international trade in conditionally cleared building and construction materials, suppliers need to verify the intended use of the materials and inform buyers of the conditions of use. It is advisable to also check the specific regulations and guidance issued by the competent authorities of the State of origin and any receiving States.

4.3. END USERS OF BUILDING AND CONSTRUCTION MATERIALS

End users (consumers) of building and construction materials include designers, builders, installers and building owners. End users are encouraged to choose and use labelled materials with clearly declared technical specifications. They are also encouraged to request the results of activity concentration measurements and any calculations, assessment reports or certificates that demonstrate the compliance of the materials with the requirements for radiation protection.

In States where an activity index or derived reference levels in terms of activity concentration are defined for building and construction materials (e.g. in building regulations or national or regional building codes), end users of building and construction materials are expected to only use materials that have been demonstrated (through appropriate testing — see Section 3.5) to be in compliance with regulations.

If the building and construction materials have been conditionally cleared, end users need be advised to only use such materials as specified in the clearance conditions.

Building and construction companies may also need to demonstrate the compliance of the final construction through dose rate measurements (see Section 3.5) and/or by providing proof that only construction materials in compliance with the national regulations and/or guidance were used.

4.4. DEMONSTRATION OF COMPLIANCE WITH REQUIREMENTS FOR BUILDING AND CONSTRUCTION MATERIALS CONTAINING RADIONUCLIDES

In order to confirm whether a building or construction material complies with requirements for radiation protection, the following information is necessary:

— Place of origin of the construction material.
— Names and addresses of supplier and producer.
— Activity concentration of all relevant radionuclides, determined by a qualified in-house laboratory or an external measurement service provider.
— A confirmation of compliance based on one or more of the following:
  • Comparison of the measured activity concentrations against the values established in regulations or guidance;
  • Results from the application of a suitable screening tool to the measured activity concentrations;
  • An assessment of the annual effective dose and comparison with the established reference level (including detailed description of calculation assumptions and parameters).
— Contact details of the qualified in-house laboratory or external measurement service provider that performed the measurements.

In order to confirm that building and construction materials in an existing building are in compliance with the reference level, the following information is needed:

— The results of dose rate measurements in different rooms or locations within a building.
— A confirmation of compliance based on one or more of the following:
  • Comparison of the dose measurement results with values established in the regulations or guidance;
  • Assessment of the annual effective dose and comparison with the established reference level (including detailed description of calculation assumptions and parameters).
— Contact details of the qualified in-house laboratory or external measurement service provider that performed the measurements.

If existing buildings need to be evaluated for compliance with the reference levels (e.g. in the case of refurbishing an industrial facility into apartments or offices, or a real estate transaction), the same list of information could be requested as above.

If the compliance of a construction material or existing building is not demonstrated, it is important that this also be adequately documented.

REFERENCES

[1] EUROPEAN COMMISSION, FOOD AND AGRICULTURE ORGANIZATION OF THE UNITED NATIONS, INTERNATIONAL ATOMIC ENERGY AGENCY, INTERNATIONAL LABOUR ORGANIZATION, OECD NUCLEAR ENERGY AGENCY, PAN AMERICAN HEALTH ORGANIZATION, UNITED NATIONS ENVIRONMENT PROGRAMME, WORLD HEALTH ORGANIZATION, Radiation Protection and Safety of Radiation Sources: International Basic Safety Standards, IAEA Safety Standards Series No. GSR Part 3, IAEA, Vienna (2014).

[2] UNITED NATIONS, Sources and Effects of Ionizing Radiation, UNSCEAR 2008 Report to the General Assembly, United Nations Scientific Committee on the Effects of Atomic Radiation (UNSCEAR), UN, New York (2010).

[3] INTERNATIONAL COMMISSION ON RADIOLOGICAL PROTECTION, The 2007 Recommendations of the International Commission on Radiological Protection, Publication 103, Elsevier, Oxford and New York (2007).

[4] EUROPEAN PARLIAMENT AND EUROPEAN COUNCIL, Regulation (EU) No. 305/2011 of 9 March 2011 Laying Down Harmonised Conditions for the Marketing of Construction Products and Repealing Council Directive 89/106/EEC, EU, Brussels (2011).

[5] INTERNATIONAL ATOMIC ENERGY AGENCY, IAEA Safety Glossary: 2018 Edition, IAEA, Vienna (2019).

[6] INTERNATIONAL COMMISSION ON RADIOLOGICAL PROTECTION, Protection of the Public in Situations of Prolonged Radiation Exposure, Publication 82, Pergamon, Oxford (1999).

[7] INTERNATIONAL ATOMIC ENERGY AGENCY, Protection of the Public Against Exposure Indoors due to Radon and Other Natural Sources of Radiation, IAEA Safety Standards Series No. SSG-32, IAEA, Vienna (2015).

[8] INTERNATIONAL ATOMIC ENERGY AGENCY, Protection of Workers Against Exposure Due to Radon, IAEA, Vienna (in preparation).

[9] SMETSERS, R., JASPER, T., A practical approach to limit the radiation dose from building materials applied in dwellings, in compliance with the European Basic Safety Standards, J. Environ. Radioact. **196** (2019) 40–49.

[10] UNITED NATIONS, Sources and Effects of Ionizing Radiation, UNSCEAR 2000 Report to the General Assembly, with Scientific Annexes, Vol. 1: Sources, United Nations Scientific Committee on the Effects of Atomic Radiation (UNSCEAR), UN, New York (2000).

[11] INTERNATIONAL ATOMIC ENERGY AGENCY, Control of Orphan Sources and Other Radioactive Material in the Metal Recycling and Production Industries, IAEA Safety Standards Series No. SSG-17, IAEA, Vienna (2012).

[12] UNITED NATIONS, Sources and Effects of Ionizing Radiation, UNSCEAR 2017 Report to the General Assembly, United Nations Scientific Committee on the Effects of Atomic Radiation (UNSCEAR), UN, New York (2017).

[13] MAKAROVSKA, O., Overview of radiological accidents involving orphan radioactive sources of ionizing radiation worldwide, Secur. Nonproliferation **2** 8 (2005) 18–25,
http://www.ntc.kiev.ua/download/arh/BTN/8.pdf

[14] INTERNATIONAL ATOMIC ENERGY AGENCY, Application of the Concepts of Exclusion, Exemption and Clearance, IAEA Safety Standards Series No. RS-G-1.7, IAEA, Vienna (2004). (A revision of this publication is in preparation.)

[15] INTERNATIONAL ATOMIC ENERGY AGENCY, Derivation of Specific Clearance Levels in Materials Being Suitable Recycling, Reuse, or for Disposal in Landfills, IAEA, Vienna (in preparation).

[16] SCHROEYERS, W. (Ed), Naturally Occurring Radioactive Materials in Construction — Integrating Radiation Protection in Reuse (COST Action Tu1301 NORM4BUILDING), Woodhead Publishing, Cambridge, UK (2017).

[17] THE RADIATION PROTECTION AUTHORITIES IN DENMARK, FINLAND, ICELAND, NORWAY AND SWEDEN, Naturally Occurring Radioactivity in the Nordic Countries — Recommendations, Swedish Radiation Protection Institute, Stockholm (2000).

[18] INTERNATIONAL ATOMIC ENERGY AGENCY, Radiation Protection Against Indoor Radon and Building and Construction Materials — Methods of Prevention and Mitigation, IAEA-TECDOC-1951, IAEA, Vienna (2021).

[19] INTERNATIONAL ATOMIC ENERGY AGENCY, Governmental, Legal and Regulatory Framework for Safety, IAEA Safety Standards Series No. GSR Part 1, IAEA, Vienna (2010).

[20] BRITISH STANDARDS INSTITUTION, BS EN 206:2013 Concrete — Specification, Performance, Production and Conformity, BSI, London (2013).

[21] CANADIAN STANDARDS ASSOCIATION, CSA A23.4-16 Precast Concrete — Materials and Construction, CSA, Toronto (2016).

[22] CANADIAN STANDARDS ASSOCIATION, CSA A23.1:19/CSA A23.2:19 Concrete Materials and Methods of Concrete Construction/Test Methods and Standard Practices for Concrete, CSA, Toronto (2019).

[23] NATIONAL RESEARCH COUNCIL CANADA, National Building Code of Canada 2015, NRCC, Ottawa (2015).

[24] BUNDESMINISTERIUMS DER JUSTIZ UND FÜR VERBRAUCHERSCHUTZ SOWIE DES BUNDESAMTS FÜR JUSTIZ, Gesetz zum Schutz vor der schädlichen Wirkung ionisierender Strahlung, BMJ, Berlin (2017),
https://www.gesetze-im-internet.de/strlschg/StrlSchG.pdf

[25] BUNDESMINISTERIUMS DER JUSTIZ UND FÜR VERBRAUCHERSCHUTZ SOWIE DES BUNDESAMTS FÜR JUSTIZ, Verordnung zum Schutz vor der schädlichen Wirkung ionisierender Strahlung, BMJ, Berlin (2018),
https://www.gesetze-im-internet.de/strlschv_2018/StrlSchV.pdf

[26] COUNCIL OF THE EUROPEAN UNION, Council Directive 2013/59/EURATOM of 5 December 2013 laying down basic safety standards for protection against the dangers arising from exposure to ionising radiation, and repealing Directives 89/618/Euratom, 90/641/Euratom, 96/29/Euratom, 97/43/Euratom and 2003/122/Euratom, EU, Brussels (2013).

[27] BRAZILIAN NUCLEAR ENERGY COMMISSION, Use of Phosphogypsum in Agriculture and the Cement Industry (Nrm488), CNEN, Rio de Janeiro (2014),
http://appasp.cnen.gov.br/seguranca/normas/pdf/Nrm488.pdf

[28] BRAZILIAN NUCLEAR ENERGY COMMISSION, Atos da Directoria de Radioproteção e Segurança Nuclear, CNEN, Rio de Janeiro (2013),
http://appasp.cnen.gov.br/seguranca/normas/pdf/pr488_01.pdf

[29] Norms of radiation safety (NRB-99/2009). Sanitary rules and norms SanPiN 2.6.1.2523-09. Approved by the resolution of the Chief state sanitary doctor of the Russian Federation of 07.07.2009 No. 47. Registered with the Ministry of Justice of the Russian Federation on Aug. 14, 2009, registration No. 14534 (2009).

[30] EUROPEAN COMMITTEE FOR STANDARDIZATION, EN 13242 — Aggregates for Unbound and Hydraulically Bound Materials for Use in Civil Engineering Work and Road Construction, CEN, Brussels (2015).

[31] AUSTRIAN STANDARDS INSTITUTE, ÖNORM EN 12620:2014 — Aggregates for Concrete, ASI, Vienna (2014).

[32] BRITISH STANDARDS INSTITUTION, BS EN 13043 — Aggregates for Bituminous Mixtures and Surface Treatments for Roads, Airfields and Other Trafficked Areas, BSI, London (2015).

[33] EUROPEAN COMMITTEE FOR STANDARDIZATION, CEN/TR 17113 Construction Products — Assessments of Releases of Dangerous Substances — Radiation from Construction Products — Dose Assessments of Emitted Gamma Radiation, CEN, Brussels (2017).

[34] INTERNATIONAL ATOMIC ENERGY AGENCY, INTERNATIONAL LABOUR ORGANIZATION, Occupational Radiation Protection, IAEA Safety Standards Series No. GSG-7, IAEA, Vienna (2018).

[35] INTERNATIONAL ATOMIC ENERGY AGENCY, Preparation and Certification of IAEA Gamma Ray Spectrometry Reference Materials RGU-1, RGTh-1 and RGK-1 (IAEA-RL-148), IAEA, Vienna (1987).

Annex I

EXAMPLES OF DOSE CALCULATION AND MODELLING

I–1. INTRODUCTION

Paragraph 5.22 of IAEA Safety Standards Series No. GSR Part 3, Radiation Protection and Safety of Radiation Sources: International Basic Safety Standards [I–1], requires the introduction of a reference level of annual effective dose due to exposure to radionuclides in commodities, including construction materials, that does not exceed a value of approximately 1 mSv/a.

To demonstrate compliance with the reference level, it is common practice to focus on proactive measures to confine the content of gamma emitting radionuclides (mainly naturally occurring) in building and construction materials before a building is constructed. Criteria for the radionuclide content of building and construction materials are derived from the reference level of effective dose using dose models. One of the dose models that can be used is the approach originally developed for use in the EU. Some basic aspects of this model are described in Section I–2 of this annex.

If the radionuclide content of building and construction materials is controlled, measurements of the gamma dose rate in buildings are considered redundant; however, if proper controls were not implemented in the previous stages, such measurements are necessary to confirm the compliance of a newly commissioned building. For this purpose, some States have established derived levels of ambient gamma dose rate to check for compliance with the reference level, either as a complementary measure to the control of building and construction materials, or as an alternative to such controls. This approach also allows for the building to be checked for contamination with artificial radionuclides, for instance as a result of melted radiation sources. Section I–3 of this annex goes into more detail on dose assessment for existing buildings.

I–2. DOSE MODELLING (EU APPROACH)

In the following, some basic principles of dose modelling according to the approach proposed by the European Committee for Standardization to be used in the EU are described.

I–2.1. Excess exposure

In accordance with the EU approach [I–2], the dose quantity for regulatory purposes is the annual excess effective dose to an occupant, referred to as 'excess exposure', E_{ex}, in the following. In this approach it is assumed that the reference level applies to the difference between the annual dose originating from building and construction materials and the dose from natural background radiation, also taking into account the attenuation of background radiation by the building to be constructed.

The consideration of the following three scenarios might be helpful in understanding what is behind the approach.

Scenario S-I: A person spends a whole year outdoors (i.e. $t = 8750$ h), being exposed to gamma radiation from the ground (terrestrial) and cosmic radiation.

Scenario S-II: A person spends a whole year indoors (i.e. $t = 8750$ h), being exposed to gamma radiation emitted by the construction and building materials; terrestrial and cosmic radiations are, however, attenuated by the building materials.

Scenario S-III: A person, considered to be representative of the population, stays indoors for part of the year (often, $t = 7000$ h is considered a plausible value, corresponding to an occupancy factor of 0.8; see below) and stays outside for the rest of the year ($t = 1750$ h).

With E_I, E_{II} and E_{III} denoting the respective annual effective doses resulting from the three scenarios above, E_{ex} is the difference between E_{III} and E_I:

$$E_{ex} = E_{III} - E_I$$

It can easily be shown that:

$$E_{ex} = Of\left(E_{BM} - a_{ter}E_{o,ter} - a_{o,cos}E_{o,cos}\right) \tag{I–1}$$

where

Of is the occupancy factor in the house (a value of 1 corresponding to scenario II, meaning that the exposure time t inside the building is 8760 h in a year);

E_{BM} is the (hypothetical) effective dose from gamma radiation only from the building material resulting from scenario II (a person living in the building all year round);

a_{ter} is an attenuation factor describing the attenuation of terrestrial exposure by the building;

a_{cos} is an attenuation factor describing the attenuation of cosmic exposure by the building;

$E_{o,ter}$ is the annual effective dose resulting from terrestrial radiation outdoors, according to scenario I;

$E_{o,cos}$ is the annual effective dose resulting from cosmic radiation outdoors, according to scenario I.

The assumptions behind this formula are as follows:

$Of = 0.9$;
$a_{ter} = 1$;
$a_{cos} = 0$;
$E_{o,ter} = 0.3$ mSv;
$\rho_{BM} = 2.3$ t/m³;
wall thickness $= 0.2$ m;
$f_k = 0.7$ (conversion factor from absorbed does rate into effective dose; see below).

These are rather conservative assumptions. For instance, a density of 2.3 t/m³ is too conservative a parameter for the vast majority of construction and building materials. Also, the assumption of a room in which all the walls, the ceiling and the floor are made from the same material with a wall thickness of 0.2 m is rather conservative.

I–2.2.   Assessment of doses and derivation of activity concentrations

As the specific usage of a construction and building material and the habits of its potential users will normally not be known at the time when the building and construction materials are to be evaluated, the models used to estimate doses use (i) standard room models (i.e. type, size, structure of the building; wall thicknesses; density of the building material and other materials) to calculate the absorbed dose rate from the relevant gamma emitting radionuclides and (ii) assumptions about a representative person with standard living habits (e.g. exposure time t; see above) and age to assess doses due to the calculated dose rate. Maximum activity concentration values for the relevant radionuclides that correspond to an exposure of 1 mSv/a can also be derived using such models.

Models to calculate absorbed dose rates from gamma emitting radionuclides in building and construction materials are usually based on Monte Carlo radiation transport modelling; see Refs [I–3, I–4] for examples.

The primary radionuclides giving rise to external radiation exposure in a building are ^{226}Ra, ^{232}Th (and their progenies) and ^{40}K.

The influence of the room size, wall thicknesses and wall densities are discussed in Ref. [I–5], which includes example calculations with varying parameters. The results are given in terms of the gamma dose rates per unit concentrations of radionuclides and the annual excess effective dose, including its dependency on the occupancy time. Exposure from superficial materials such as tiles is discussed, as are cases in which only part of the room is made of the building and construction materials (e.g. the walls).

Using this model and a reference level of 1 mSv/a, maximum concentrations of the relevant radionuclides were derived, and an activity index (denoted AI in equation (I–2) below) was defined to take into account the summing up of the three contributions:

$$AI = \frac{C_{Ra}}{300} + \frac{C_{Th}}{200} + \frac{C_K}{3000} \tag{I–2}$$

The assumptions behind this formula are similar to those relating to formula (I–1). The activity index formula[1] (I–2) is, therefore, generally considered to be a conservative screening tool. It will be possible to show in many cases (if not the majority of them) that the reference level is not exceeded if more realistic parameters are used, such as those considered in the example calculations given in Ref. [I–5]. Based on such calculations, the definition of a modified activity index using different maximum concentration levels is also possible, if this seems appropriate for the individual State; see Ref. [I–6].

If compliance with the activity index (however defined in a State) cannot be demonstrated, the next step would normally be dose assessments using more realistic assumptions for the parameters, such as the density of the building and construction material or its use in a building (e.g. in various layers).

An example of how this can be done is the methodology developed by the European Committee for Standardization (CEN) [I–7]. Although important assumptions and parameters in the calculation, such as exposure time, building geometry and terrestrial background, remain the same, the methodology results in more realistic dose assessments by providing the opportunity to take into account

(a) Realistic material densities and wall thicknesses;
(b) Different materials used for walls, ceilings or floors;
(c) Multiple layers of materials and composite materials.

This is an example in which, for the sake of consistency, certain parameters need to always remain the same within the boundaries of a given regulation, regardless of whether they are used in a general manner to derive an activity

---

[1] This formula forms the basis of the regulations on building and construction materials in the EU's basic safety standards [I–2].

index as a screening tool or for detailed dose calculations in a specific application. This is especially important for the parameters characterizing the representative person (e.g. occupancy times) or, as far as relevant, the background deduction in this approach.

I–3. DOSE ASSESSMENT FOR EXISTING BUILDINGS

It has already been stressed that control of exposure from building and construction materials is normally carried out by regulating their activity concentration prior to the construction of a building. If the building or construction material complies with either the activity index or the reference level, a final check of the resulting exposure to the occupants of a building is normally not necessary. However, in the absence of such controls on building and construction materials, measurements of the gamma dose rate may be needed to evaluate the situation. In some States, these gamma dose rate measurements are a mandatory part of acceptance tests performed before commissioning the building [I–6].

A number of States have chosen to set derived reference levels for gamma dose rates in buildings. Reference [I–6] provides an overview of how different States derive reference levels for dose rates in existing buildings. The overview shows that values vary between 0.2 µSv/h in countries such as Belarus, Bulgaria and Georgia and 0.5 µSv/h in countries such as Estonia and Slovakia.

Using the results of dose rate measurements, the exposure from external radiation in a building is calculated using the following relation:

$$E_{\text{building}} = \dot{H}^*(10)tf_k \tag{I–3}$$

where

$\dot{H}^*(10)$ is the ambient dose equivalent (e.g. in mSv/h);
t is the occupancy hours (h);

and $f_k = 0.7$ (conversion factor from ambient dose equivalent into effective dose; see below).

Care is needed when trying to compare the results of such measurements and resulting doses with the results of the modelling described above. One of the challenges is the differentiation between the contribution of natural background radiation outside of the measured premises and other surrounding buildings to the measured dose rate, especially when the shielding effect of the building material is taken into account.

I–4. ARTIFICIAL RADIONUCLIDES

Normally, artificial radionuclides in building and construction materials will not give rise to exposures that exceed the reference level. No generally applicable models and tools for artificial radionuclides have been derived to check consistency with the reference level.

The only relevant model seems to be the one used to derive exemption and clearance levels in Ref. [I–8]. The derived activity concentrations for artificial radionuclides are based on an annual effective dose of the order of a few tens of microsieverts per year and are not applicable for the evaluation of building and construction materials without further consideration.

In the event that an accidental occurrence of a source or contamination of a construction or building material is discovered, it is likely that the dose calculation will be based directly on dose rate measurements; see Section I–3.

There are known cases of contaminated materials used in construction, such as steel that was contaminated with ^{60}Co by accidentally melting a large source in a steel factory. The reference level of 1 mSv/a is valid for exposure from all radionuclides and thus valid for such situations as well. Specific calculation methods are used to calculate doses in such cases, e.g. the method found in Ref. [I–4].

REFERENCES TO ANNEX I

[I–1] EUROPEAN COMMISSION, FOOD AND AGRICULTURE ORGANIZATION OF THE UNITED NATIONS, INTERNATIONAL ATOMIC ENERGY AGENCY, INTERNATIONAL LABOUR ORGANIZATION, OECD NUCLEAR ENERGY AGENCY, PAN AMERICAN HEALTH ORGANIZATION, UNITED NATIONS ENVIRONMENT PROGRAMME, WORLD HEALTH ORGANIZATION, Radiation Protection and Safety of Radiation Sources: International Basic Safety Standards, IAEA Safety Standards Series No. GSR Part 3, IAEA, Vienna (2014).

[I–2] COUNCIL OF THE EUROPEAN UNION, Council Directive 2013/59/EURATOM of 5 December 2013 laying down basic safety standards for protection against the dangers arising from exposure to ionising radiation, and repealing Directives 89/618/Euratom, 90/641/Euratom, 96/29/Euratom, 97/43/Euratom and 2003/122/Euratom, EU, Brussels (2013).

[I–3] BERGER, M.J., Calculation of energy dissipation by gamma radiation near the interface between two media, J. Appl. Phys. **28** 12 (1957) 1502–1508.

[I–4] MERK, R., et al., PENELOPE-2008 Monte Carlo simulation of gamma exposure induced by ^{60}Co and NORM-radionuclides in closed geometries, Appl. Radiat. Isot. **82** (2013) 20–27.

[I–5] INTERNATIONAL ATOMIC ENERGY AGENCY, WORLD HEALTH ORGANIZATION, Protection of the Public Against Exposure Indoors due to Radon and Other Natural Sources of Radiation, IAEA Safety Standards Series No. SSG-32, IAEA, Vienna (2015).

[I–6] INTERNATIONAL ATOMIC ENERGY AGENCY, Status of Radon Related Activities in Member States Participating in Technical Cooperation Projects in Europe, IAEA-TECDOC-1810, IAEA, Vienna (2017).

[I–7] INTERNATIONAL ATOMIC ENERGY AGENCY, Protection of the Public Against Exposure Indoors due to Radon and Other Natural Sources of Radiation, IAEA Safety Standards Series No. SSG-32, IAEA, Vienna (2015).

[I–8] INTERNATIONAL ATOMIC ENERGY AGENCY, Derivation of Specific Clearance Levels in Materials Being Suitable Recycling, Reuse, or for Disposal in Landfills, IAEA Safety Report, IAEA, Vienna (in preparation).

Annex II

EXAMPLES OF MEASUREMENT METHODS AND INSTRUMENTS FOR THE DETERMINATION OF RADIOACTIVITY IN AND DOSE RATES FROM BUILDING AND CONSTRUCTION MATERIALS

II–1. INTRODUCTION

In this annex, a general overview of different measurement instruments and methods for the determination of radioactivity in building and construction materials will be given. A number of options and possibilities are introduced, and examples of methods that are already in use in some States are summarized. This section addresses the determination of the activity concentration inside a building and construction material as well as dose rate measurements.

II–2. MEASUREMENT INSTRUMENTS

A variety of gamma spectrometry equipment may be used for the determination of the activity concentration of different radionuclides in a material. High purity germanium (HPGe) and sodium iodide (NaI) detectors are the most widely used types of detector for this purpose. However, other detector types, such as bismuth germanium oxide (BGO) and cadmium zinc based (CdZnTe or CdZn) detectors, are becoming more popular. The methods described in the examples below show how different detector systems may be used to demonstrate compliance and/or to screen building and construction materials.

For the purpose of dose rate measurements, a variety of dose rate meters with different detection ranges and sensitivities are available on the market. The choice of dose rate meter will depend on the purpose of the measurements.

II–3. SWEDISH MMK METHODS

In Sweden, the testing of dangerous substances, including radioactivity, in stone based building materials is performed by the producers to comply with EU Construction Product Regulation No. 305/2011 and the new Swedish Radiation Protection Act (2018:396) [II–1]. To comply with this, accredited testing laboratory Mark-och Miljokontroll (MMK) has developed a measurement methodology entitled Method MMK A2 605, Determination of Activity Index, Radium Index and Gamma Radiation in Stone Based Construction Material, for testing building and construction materials such as concrete and ballast (crushed stone) products [II–2].

The method allows for the activity concentration of gamma emitting radionuclides in the construction material to be determined using gamma spectrometry. The activity concentration is measured in a sample of 150 mm × 150 mm × 150 mm. A detector (BGO crystal) with a volume exceeding 100 cm³ and shielded with steel covered lead, for example, is used. This determination of activity concentration in a sample reflects the building material in the intended form of use. The calculation of the activity index, radium index and gamma radiation is performed according to the formulas below. The activity index is calculated as follows:

$$AI = \frac{C_{Ra} \times 12.35}{300} + \frac{C_{Th} \times 4.06}{200} + \frac{C_{K} \times 313}{3000} \tag{II–1}$$

where

C_K is the content of potassium in percent;
C_{Ra} is the content of radium (^{238}U or ^{235}U) in ppm;

and C_{Th} is the content of thorium in ppm.

The radium index is calculated as follows:

$$RI = \frac{C_{Ra} \times 12.35}{300} \tag{II–2}$$

where

RI is the radium index;

and C_{Ra} is the content of radium (^{238}U or ^{235}U) in ppm.

The gamma dose is calculated, on a flat surface 2π measurement, as follows:

$$E = \left(C_K \times 0.0151\right) + \left(C_{Ra} \times 0.0065\right) + \left(C_{Th} \times 0.0029\right) \tag{II–3}$$

where

E is the effective dose;
C_K is the content of potassium in percent;
C_{Ra} is the content of radium (uranium 238/235) in ppm;

and C_{Th} is the content of thorium in ppm.

Two methods are used for the determination of the annual effective dose in existing buildings.

The Swedish method MMK 606, In-Situ Method for Determination of Activity Index (AI) of Construction Materials Indoors of Existing Buildings, is for the measurement of the activity concentration of NORM in an existing building and calculation of the annual effective dose [II–3]. The activity concentration of gamma emitting radionuclides in building and construction materials is determined by gamma spectrometry based on the same measurement instrument principle as the method above (BGO crystal with a volume exceeding 100 cm³ and shielding with, for example, steel covered lead). The activity concentration is measured in a room directly on a surface of the building material. The activity index is calculated according to (II–1) and the effective dose according to (II–3).

In cases where gamma spectrometric measurements are not possible or not necessary, MMK 608, Determination of Effective Dose — Gamma Radiation from Stone Based Construction Material, can be used [II–4]. This method utilizes a detector (NaI (Tl) crystal) with a volume exceeding 345 cm³ (diameter 76 × 76 mm). The dose rate is measured in the air in a room of a building. Effective dose is calculated using the following formula:

$$E_{eff} = E \times t \tag{II–4}$$

where

E_{eff} is the annual effective dose in mSv/a;

and t is the exposure time for residential housing (7000 h) or for workplaces (2000 h).

The three methods above allow rapid and reliable results to be obtained and can be used by authorities as well as manufacturers, suppliers and others for verification of compliance of building and construction materials or existing buildings.

II–4. CEN/TS 17216 METHOD

Another example of a measurement method is CEN/TS 17216. In 2018, the European Committee for Standardization (CEN) published a Technical Specification entitled Determination of Activity Concentrations of Radium-226, Thorium-232 and Potassium-40 in Construction Products using Semiconductor Gamma-Ray Spectrometry [II–5], which is currently being reviewed and tested to determine whether it can be converted into an official European Standard (EN standard). The information on the radioactivity content of construction products given a Conformité Européenne (CE) mark will be based upon measurements made in accordance with this standard (CE marking is a certification mark that indicates conformity with health, safety and environmental protection standards for products sold within the European Economic Area).

In this method, high-efficiency semiconductor gamma ray spectrometers (such as HPGe detectors) are used to determine ^{226}Ra, ^{232}Th and ^{40}K. Radium and thorium are determined in an indirect manner through their daughter products ^{214}Pb and ^{228}Ac. However, ^{226}Ra decays into ^{214}Pb via ^{222}Rn, and as radon is a noble gas, it may escape from the sample before it decays into its daughter products, which would cause the activity concentration of ^{226}Ra to be underestimated. For this reason, all samples are required to be sealed in a radon-tight container. Furthermore, to ensure secular equilibrium between ^{226}Ra and ^{214}Pb in the sealed samples, they are measured after a waiting period of around three weeks.

Although this measurement method may provide accurate results and a lower limit of detection, its primary drawback is the waiting time needed before measurement and the lack of a calibration standard material. This waiting period may not be an issue for CE certification; however, when radioactivity levels need to be determined quickly in the industrial process, a three week wait will pose a problem. This may apply, for example, in cases where batches of materials in quarries or mines need to be screened quickly before transport and further processing. For these purposes, a screening method with a short measurement time may be applied, such as the Swedish MMK A2 605 method outlined in Section II–3.

II–5. DETERMINATION OF DOSE RATE

There are numerous different types of dose rate meters available for the purpose of determining the dose rate inside an existing building. The dose calculation models and their corresponding parameters that are explained in Annex I need to be taken into account when a dose rate meter is chosen for this application. For example, considering the most widely used occupancy factor for dwellings of 7000 h and a reference level of 1 mSv/a, dose rates of the order of 100 ηSv/h need to be measured.

To ensure that measurement results are comparable, dose rate measurements always need to be carried out in a similar position relative to the geometry of the room. For example, it is good practice to always choose a position in the middle of a room, or at least 1 m away from a wall, where possible.

REFERENCES TO ANNEX II

[II–1] EUROPEAN PARLIAMENT AND EUROPEAN COUNCIL, Regulation (EU) No. 305/2011 of the European Parliament and of the Council of 9 March 2011 laying down harmonised conditions for the marketing of construction products and repealing Council Directive 89/106/EEC, EU, Brussels (2011).

[II–2] LÖFQVIST, T., LÖFQVIST, G., GERMAN O., Determination of activity index, radium index and gamma radiation in stone-based construction material, MMK A2 605, Mark- och Miljökontroll i Särö AB, Billdal, Sweden (2016), https://markochmiljokontroll.se/en/method-mmk-605/

[II–3] LÖFQVIST, T., LÖFQVIST, G., GERMAN, O., In-situ method for determination of Activity Index (AI) of construction materials indoors of existing buildings, MMK A2 606, Mark- och Miljökontroll i Särö AB, Billdal, Sweden (2016), https://markochmiljokontroll.se/en/method-mmk-606/

[II–4] LÖFQVIST, T., LÖFQVIST, G., Determination of effective dose — Gamma Radiation from stone-based Construction material, MMK 608, Mark- och Miljökontroll i Särö AB, Billdal, Sweden (2020), https://markochmiljokontroll.se/en/method-mmk-608/

[II–5] EUROPEAN COMMITTEE FOR STANDARDIZATION, Construction products — Assessment of release of dangerous substances — Determination of activity concentrations of radium-226, thorium-232 and potassium-40 in construction products using semiconductor gamma-ray spectrometry, CEN/TS 17216:2018, CEN, Brussels (2018).

Annex III

INFORMATION ABOUT BUILDING AND CONSTRUCTION MATERIALS ENTERING THE MARKET

III–1. INTRODUCTION

The authorities in different States publish information relating to the radioactivity in commonly used building and construction materials, including natural stone, gravel, fly ash, and reused and recycled materials. NORM residues from various industrial processes, such as the processes used in fossil fuel extraction, processing and power plants, the phosphate industry, the aluminium industry and other metal industries, are commonly used in many States and are partly included in respective overviews.

Information on the typical content in different materials of naturally occurring radionuclides, including their variability, may help both radiation protection experts not normally engaged with natural radioactivity and non-experts from concerned industries or from the general public to at least roughly assess the possible radiological meaning of the materials.

In Tables III–1 to III–3, examples of information gathered and published by the State authorities in Germany, the Czech Republic and China, respectively, are reproduced.

III–2. GERMANY

In Germany, the Federal Office for Radiation Protection has published data on the activity concentrations of ^{226}Ra, ^{232}Th and ^{40}K in a variety of materials that are, or might be, of importance as building and construction materials, either directly or as constituents [III–1].

III–3. THE CZECH REPUBLIC

The National Radiation Protection Institute of the Czech Republic lists the average and maximum activity concentrations of ^{226}Ra in the building and construction materials listed in Table III–2 [III–2].

TABLE III–1. ACTIVITY CONCENTRATIONS OF NATURAL RADIONUCLIDES IN STONES, BUILDING AND CONSTRUCTION MATERIALS, AND NORM RESIDUES IN GERMANY

(adapted with permission from Ref. [III–1])

Material	Ra-226, Bq/kg Mean value (range)	Th-232, Bq/kg Mean value (range)	K-40, Bq/kg Mean value (range)
Granite	100 (30–500)	120 (17–311)	1000 (600–4000)
Gneiss	75 (50–157)	43 (22–50)	900 (830–1500)
Diabase	16 (10–25)	8 (4–12)	170 (100–210)
Basalt	26 (6–36)	29 (9–37)	270 (190–380)
Granite	10 (4–16)	6 (2–11)	360 (9–730)
Gravel, sand, gravel sand	15 (1–39)	16 (1–64)	380 (3–1200)
Natural gypsum, anhydrite	10 (2–70)	5 (2–100)	60 (7–200)
Tuff, pumice stone	100 (<20–200)	100 (30–300)	1000 (500–2000)
Clay	<40 (<20–90)	60 (18–200)	1000 (300–2000)
Brick, clinker brick	50 (10–200)	52 (12–200)	700 (100–2000)
Concrete	30 (7–92)	23 (4–71)	450 (50–1300)
Sand-lime brick, porous concrete	15 (6–80)	10 (1–60)	200 (40–800)
Slag from copper production	1500 (860–2100)	48 (18–78)	520 (300–730)
Gypsum from flue gas desulphurization	20 (<20–70)	<20	<20
Brown coal filter ash	82 (4–200)	51 (6–150)	147 (12–610)

TABLE III–2. RADIUM-226 ACTIVITY CONCENTRATIONS IN
BUILDING AND CONSTRUCTION MATERIALS IN THE CZECH
REPUBLIC
(adapted with permission from Ref. [III–2])

Building material	Ra-226, Bq/kg Mean value (range)	Ra-226, Bq/kg Maximum value (range)
Building stone	27.5	925
Bricks	45.2	143
Concrete	21.1	192
Porous concrete	46.1	85
Clinker concrete	66.7	118
Mortars	19.8	82
Plasters	13.9	56
Ceramic tiles	63.0	117
Sand	13.3	41
Clay	40.9	199
Aggregate	34.9	1090
Fly ash and slag	75.5	363
Cement	36.5	88
Lime	12.5	94
Gypsum	12.1	86

III–4. CHINA

Reference [III–3] presents an overview of activity concentrations of
^{235}U, ^{238}U, ^{232}Th, ^{226}Ra, ^{210}Pb and ^{40}K measured in five types of typical general
industrial solid waste in Guizhou, China, including fly ash, red mud, phosphorus
slag, phosphogypsum and electrolytic manganese residue. The radiation levels
of phosphogypsum, electrolytic manganese residue and electrolytic manganese
residue activated by NaOH or $Ca(OH)_2$ were all lower than the nationally

occurring levels in China. The values of the internal and external radiation index (I_{Ra} and I_γ, respectively) for fly ash and red mud were higher ($I_{Ra} > 1.0$ and $I_\gamma > 1.3$ for fly ash, $I_{Ra} > 2.0$ and $I_\gamma > 2.0$ for red mud) (see Annex IV). Reference [III–3] further recommends that construction material should not contain more than 75.44%, 29.72% and 66.01% of the tested red mud, fly ash and phosphorus slag, respectively, from a radiation protection perspective.

Reference [III–3] also summarizes the average activity concentrations of ^{226}Ra, ^{232}Th and ^{40}K radionuclides in fly ash, red mud, phosphorus slag, phosphogypsum and electrolytic manganese residue samples from other countries as shown in similar studies (see Table III–3 below).

TABLE III–3. AVERAGE ACTIVITY CONCENTRATION OF Ra-226, Th-232 AND K-40 IN FLY ASH, RED MUD, PHOSPHOGYPSUM AND ELECTROLYTIC MANGANESE RESIDUE SAMPLES FROM STUDIES IN DIFFERENT PARTS OF THE WORLD
(adapted with permission from Ref. [III–3])

Country	Sample type	Activity concentration (Bq/kg)		
		Ra-226	Th-232	K-40
India		119	147	352
Türkiye		360	102	517
Hungary		178	55	387
Greece		815	56	400
Czech Republic		146	8	669
Germany	Fly ash	164	94	517
Italy		170	140	400
Poland		200	118	798
Romania		219	116	595

TABLE III–3. AVERAGE ACTIVITY CONCENTRATION OF Ra-226, Th-232 AND K-40 IN FLY ASH, RED MUD, PHOSPHOGYPSUM AND ELECTROLYTIC MANGANESE RESIDUE SAMPLES FROM STUDIES IN DIFFERENT PARTS OF THE WORLD
(adapted with permission from Ref. [III–3]) (cont.)

Country	Sample type	Activity concentration (Bq/kg)		
		Ra-226	Th-232	K-40
China (Baoji)		112	148	386
China (Xiangyang)		441	110	510
Türkiye		210	539	112
Hungary		301	295	50
Greece		244	364	57
Germany	Red mud	171	318	215
Italy		97	118	115
Australia		310	1350	350
Jamaica		1047	350	335
Türkiye		436	9	13
Bangladesh		234	21	108
Egypt		596	6	2
Jordan	Phosphogypsum	376	4	40
Republic of Korea		618	9	24
Israel		747	14	63

TABLE III–3. AVERAGE ACTIVITY CONCENTRATION OF Ra-226,
Th-232 AND K-40 IN FLY ASH, RED MUD, PHOSPHOGYPSUM AND
ELECTROLYTIC MANGANESE RESIDUE SAMPLES FROM STUDIES IN
DIFFERENT PARTS OF THE WORLD
(adapted with permission from Ref. [III–3]) (cont.)

| Country | Sample type | Activity concentration (Bq/kg) | | |
		Ra-226	Th-232	K-40
Spain		647	8	33
South Africa		109	189	>100
Hungary	Electrolytic manganese residue	52	40	607
China (Chongqing)		37	58	631

III–5. SPAIN

The Nuclear Safety Council (CSN), the competent authority for nuclear safety and radiation protection in Spain, has supervised a wide testing programme to assess the radiological risk associated with white and grey Portland cements [III–4]. The sampling procedure was developed by the Technical Institute of the Spanish Cement Producers Association, and the activity concentration determined using gamma spectrometry was performed by the department for environmental radioactivity and the radiological surveillance unit for the Radiation Protection of the Public and the Environment group of the Centre for Energy, Environmental and Technology Research (CIEMAT).

The activity concentrations of ^{226}Ra, ^{232}Th and ^{40}K were determined in 11 white cement samples from 4 Spanish factories and 63 grey Portland cement samples from 26 Spanish factories. Furthermore, 79 coal fly ash and natural pozzolan Portland cement samples were provided by 22 Spanish integrated cement plants and 3 cement grinding plants from 12 different regions of Spain. In total, 67 coal fly ash and 13 natural pozzolan Portland cement samples were collected.

Table III–4 shows the minimum, maximum and average activity concentrations of ^{226}Ra, ^{232}Th and ^{40}K, and the activity index, *AI*, obtained in white and grey cement samples [III–4]. The activity index for all the Portland cements without additions is much lower than the reference level of the unity.

TABLE III–4. ACTIVITY CONCENTRATIONS OF Ra-226, Th-232 AND K-40 (BQ/KG), AND ACTIVITY INDEX, *AI*, OF WHITE AND GREY CEMENT SAMPLES
(adapted with permission from Ref. [III–4])

White cement samples	Ra-226	Th-232	K-40	*AI*
Maximum	22.55 ± 0.88	16.1 ± 4.8	239 ± 13	0.1929 ± 0.0045
Minimum	5.61 ± 0.36	5.20 ± 0.27	41.8 ± 2.7	0.0586 ± 0.0020
Average	13.5 ± 2.1	8.52 ± 0.84	133 ± 24	0.132 ± 0.017
World average	26.7	17.4	93.1	0.2
Grey cement samples	Ra-226	Th-232	K-40	*AI*
Maximum	59.4 ± 4.6	20.1 ± 1.8	310 ± 13	0.379 ± 0.019
Minimum	10.09 ± 0.57	4.97 ± 0.37	60.5 ± 5.3	0.0797 ± 0.0036
Average	24.4 ± 1.3	13.09 ± 0.41	181.7 ± 8.5	0.2074 ± 0.0079

REFERENCES TO ANNEX III

[III–1] BUNDESAMT FÜR STRAHLENSCHUTZ, Radionuclides in building materials, BfS, Salzgitter, Germany (2022), https://www.bfs.de/EN/topics/ion/environment/building-materials/radionuclides/radionuclides_node.html

[III–2] CZECH NATIONAL RADIATION PROTECTION INSTITUTE (SÚRO), Building materials, SÚRO, Prague (2022), https://www.suro.cz/en/prirodnioz/building-materials

[III–3] SHEN, Z., ZHANG, Q., CHENG, W., CHEN, Q., Radioactivity of five typical general industrial solid wastes and its influence in solid waste recycling, Minerals **9** (2019) 168.

[III–4] SANJUÁN, M.Á., SUAREZ-NAVARRO, J.A., ARGIZ, C., MORA, P., Assessment of radiation hazards of white and grey Portland cements. J. Radioanal. Nucl. Chem. **322** 2 (2019) 1169–1177.

Annex IV

EXAMPLES OF NATIONAL REGULATIONS ON RADIONUCLIDES IN BUILDING AND CONSTRUCTION MATERIALS

IV–1. INTRODUCTION

Paragraph 5.22 of IAEA Safety Standards Series No. GSR Part 3, Radiation Protection and Safety of Radiation Sources: International Basic Safety Standards [IV–1], requires the introduction of a reference level of annual effective dose due to exposure to radionuclides in commodities, including construction materials, that does not exceed a value of approximately 1 mSv/a. Member States have different approaches to implementing this reference level in their national legal and regulatory systems, in terms of the legal instruments used as well as the choice of the most appropriate body for the task of regulatory control (e.g. the regulatory body for safety, building and construction safety authorities or public health safety authorities).

IV–2. EXAMPLES FROM MEMBER STATES

Some examples of how some Member States implemented this requirement as of the end of 2016 are provided below, based on Refs [IV–2, IV–3] and information provided directly by Member States. Table IV-1 provides a summary of legislation and recommendations for the regulatory of radionuclides in building and construction materials.

Text cont. on p. 62.

TABLE IV–1. SUMMARY OF THE LEGISLATION AND RECOMMENDATIONS USED IN SOME MEMBER STATES FOR THE REGULATORY CONTROL OF RADIONUCLIDES IN BUILDING AND CONSTRUCTION MATERIALS

Country	Activity index	Regulation of Ra-226 content?	Decision values	Reference levels and gamma dose rate criteria
Albania (Government of Albania, 2011)	$AI_1 = \dfrac{C_{Ra}}{300} + \dfrac{C_{Th}}{200} + \dfrac{C_K}{3000}$ $AI_2 = \dfrac{C_{Ra}}{700} + \dfrac{C_{Th}}{500} + \dfrac{C_K}{8000} + \dfrac{C_{Cs}}{2000}$	No	$AI_1 \leq 1$ Bulk material $AI_2 \leq 1$ Bulk material for road construction	1 mSv/a
Argentina [IV–4]				1 mSv/a
Armenia	$C_{eff} = C_{Ra} + 1.30\, C_{Th} + 0.09\, C_k$	No	$C_{eff} \leq 370$ Bq/kg If $C_{eff} > 4000$ Bq/kg the materials should not be used in buildings	0.3 µSv/h reference level 0.6 µSv/h limit
Austria	$AI = \dfrac{C_{Ra}}{880}(1 + 0.07\varepsilon \times d \times p) + \dfrac{C_{Th}}{530} + \dfrac{C_K}{8800}$	No	$AI \leq 1$	1 mSv/a

TABLE IV–1. SUMMARY OF THE LEGISLATION AND RECOMMENDATIONS USED IN SOME MEMBER STATES FOR THE REGULATORY CONTROL OF RADIONUCLIDES IN BUILDING AND CONSTRUCTION MATERIALS (cont.)

Country	Activity index	Regulation of Ra-226 content?	Decision values	Reference levels and gamma dose rate criteria
Belarus	$C_{eff} = C_{Ra} + 1.30\, C_{Th} + 0.09\, C_K$	No	$C_{eff} \leq 370$ Bq/kg If $C_{eff} > 4000$ Bq/kg the materials should not be used in buildings	Reference level of 0.2 µSv/h
Belgium	$AI = \dfrac{C_{Ra}}{300} + \dfrac{C_{Th}}{200} + \dfrac{C_K}{3000}$			1 mSv/a
Brazil [IV–5]		Yes[a]	Exemption level for C_{Ra-226} or C_{Ra-228} ≤ 1000 Bq/kg	
Bulgaria		No	$I \leq 1$	Reference level of 0.2 µSv/h

TABLE IV–1. SUMMARY OF THE LEGISLATION AND RECOMMENDATIONS USED IN SOME MEMBER STATES FOR THE REGULATORY CONTROL OF RADIONUCLIDES IN BUILDING AND CONSTRUCTION MATERIALS (cont.)

Country	Activity index	Regulation of Ra-226 content?	Decision values	Reference levels and gamma dose rate criteria
China	$AI_r = \dfrac{C_{Ra}}{370} + \dfrac{C_{Th}}{260} + \dfrac{C_K}{4200}$; $I_{Ra} = \dfrac{C_{Ra}}{200}$	Yes	Bulk material internal exposure index $I_{Ra} \leq 1$, external exposure index $I_r \leq 1$ Superficial and 25% hollow bulk material for dwelling construction $I_{Ra} \leq 1$, $AI_r \leq 1.3$ Superficial material for industrial construction $I_{Ra} \leq 1.3$, $AI_r \leq 1.9$ Superficial material for outside use $AI_r \leq 2.8$	1 mSv/a
Croatia	$AI = \dfrac{C_{Ra}}{300} + \dfrac{C_{Th}}{200} + \dfrac{C_K}{3000}$	No	$AI \leq 1$	

53

TABLE IV–1. SUMMARY OF THE LEGISLATION AND RECOMMENDATIONS USED IN SOME MEMBER STATES FOR THE REGULATORY CONTROL OF RADIONUCLIDES IN BUILDING AND CONSTRUCTION MATERIALS (cont.)

Country	Activity index	Regulation of Ra-226 content?	Decision values	Reference levels and gamma dose rate criteria
Cyprus	$AI = \dfrac{C_{Ra}}{300} + \dfrac{C_{Th}}{200} + \dfrac{C_K}{3000}$	No	$AI \leq 1$	1 mSv/a
Czech Republic	$AI = \dfrac{C_{Ra}}{300} + \dfrac{C_{Th}}{200} + \dfrac{C_K}{3000}$	Yes	Bulk material $AI \leq 0.5$ $C_{Ra} \leq 150$ Bq/kg	0.5 µSv/h for new buildings and a limit of 1 µSv/h for existing buildings
			Raw material $AI \leq 1$ $C_{Ra} \leq 300$ Bq/kg	0.3 mSv/a
			Superficial material $AI \leq 2$ $C_{Ra} \leq 300$ Bq/kg	

TABLE IV–1. SUMMARY OF THE LEGISLATION AND RECOMMENDATIONS USED IN SOME MEMBER STATES FOR THE REGULATORY CONTROL OF RADIONUCLIDES IN BUILDING AND CONSTRUCTION MATERIALS (cont.)

Country	Activity index	Regulation of Ra-226 content?	Decision values	Reference levels and gamma dose rate criteria
Estonia	$AI = \dfrac{C_{Ra}}{300} + \dfrac{C_{Th}}{200} + \dfrac{C_K}{3000}$	No	$AI \leq 1$	A limit of 0.5 µSv/h for schools and preschools
Finland	$AI_1 = \dfrac{C_{Ra}}{300} + \dfrac{C_{Th}}{200} + \dfrac{C_K}{3000}$ $AI_2 = \dfrac{C_{Ra}}{700} + \dfrac{C_{Th}}{500} + \dfrac{C_K}{8000} + \dfrac{C_{Cs}}{2000}$	No	Bulk material $AI_1 \leq 1$ Superficial material $AI_1 \leq 6$ Bulk material for road construction $AI_2 \leq 1$ Superficial material for road construction $AI_2 \leq 1.5$	1 mSv/a 0.1 mSv/a
Georgia		No		A reference level of 0.2 µSv/h

TABLE IV–1. SUMMARY OF THE LEGISLATION AND RECOMMENDATIONS USED IN SOME MEMBER STATES FOR THE REGULATORY CONTROL OF RADIONUCLIDES IN BUILDING AND CONSTRUCTION MATERIALS (cont.)

Country	Activity index	Regulation of Ra-226 content?	Decision values	Reference levels and gamma dose rate criteria
Greece	$AI = \dfrac{C_{Ra}}{300} + \dfrac{C_{Th}}{200} + \dfrac{C_K}{3000}$	No	$AI \leq 1$	
Kazakhstan	$C_{eff} = C_{Ra} + 1.30\,C_{Th} + 0.09\,C_k$	No	$C_{eff} \leq 370$ Bq/kg	
Latvia		Yes	Action level: $C_{Ra} \sim 100$ Bq/kg Limit value of C_{Ra}, $C_{Th} \leq 270$ Bq/kg Reference level for sources of gamma radiation: 1 Bq/kg Limit value for sources of gamma radiation: 2 Bq/kg	A limit of 0.5 μSv/h

TABLE IV–1. SUMMARY OF THE LEGISLATION AND RECOMMENDATIONS USED IN SOME MEMBER STATES FOR THE REGULATORY CONTROL OF RADIONUCLIDES IN BUILDING AND CONSTRUCTION MATERIALS (cont.)

Country	Activity index	Regulation of Ra-226 content?	Decision values	Reference levels and gamma dose rate criteria
Lithuania	$AI = \dfrac{C_{Ra}}{300} + \dfrac{C_{Th}}{200} + \dfrac{C_K}{3000}$	No	$AI \leq 1$	A reference level of 0.35 µSv/h for dwellings and a reference level of 0.45 µSv/h for workplaces
Montenegro	$AI = \dfrac{C_{Ra}}{300} + \dfrac{C_{Th}}{200} + \dfrac{C_K}{3000}$	No	$AI \leq 1$	
North Macedonia	$AI = \dfrac{C_{Ra}}{300} + \dfrac{C_{Th}}{200} + \dfrac{C_K}{3000}$	No	$AI < 0.5$ for unrestricted use of building materials $0.5 < AI < 2$ for limited use of building materials	Action level: 0.3 mSv/a

TABLE IV–1. SUMMARY OF THE LEGISLATION AND RECOMMENDATIONS USED IN SOME MEMBER STATES FOR THE REGULATORY CONTROL OF RADIONUCLIDES IN BUILDING AND CONSTRUCTION MATERIALS (cont.)

Country	Activity index	Regulation of Ra-226 content?	Decision values	Reference levels and gamma dose rate criteria
Poland [IV–2]			$C_{Ra} \leq 276 \ Bq/m^3$ $C_{Th} \leq 231 \ Bq/m^3$ $C_K \leq 3716 \ Bq/m^3$	
Republic of Moldova	$C_{eff} = C_{Ra} + 1.31 \ C_{Th} + 0.09 \ C_K$	No	$C_{eff} \leq 300 \ Bq/kg$	A reference level of 0.25 μSv/h and a limit of 0.5 μSv/h
Romania	$AI = \dfrac{C_{Ra}}{300} + \dfrac{C_{Th}}{200} + \dfrac{C_K}{3000}$	No	Legal framework contains a list of building materials with admitted mass activities, otherwise $AI \leq 0.5$	

TABLE IV–1. SUMMARY OF THE LEGISLATION AND RECOMMENDATIONS USED IN SOME MEMBER STATES FOR THE REGULATORY CONTROL OF RADIONUCLIDES IN BUILDING AND CONSTRUCTION MATERIALS (cont.)

Country	Activity index	Regulation of Ra-226 content?	Decision values	Reference levels and gamma dose rate criteria
Russian Federation	$C_{eff} = C_{Ra} + 1.30\, C_{Th} + 0.09\, C_K$	No	General: $C_{eff} \leq 370$ Industrial buildings and facilities, in road construction in areas of prospective development: $370 \leq C_{eff} \leq 740$ Road construction: $740 \leq C_{eff} \leq 1500$ The use decided by the federal executive authority that is responsible for federal state sanitary and epidemiological supervision: $1500 \leq C_{eff} \leq 4000$ Not to be used in construction: $C_{eff} > 4000$	A limit of 0.3 µSv/h for dwellings and a limit of 0.6 µSv/h for workplaces

TABLE IV–1. SUMMARY OF THE LEGISLATION AND RECOMMENDATIONS USED IN SOME MEMBER STATES FOR THE REGULATORY CONTROL OF RADIONUCLIDES IN BUILDING AND CONSTRUCTION MATERIALS (cont.)

Country	Activity index	Regulation of Ra-226 content?	Decision values	Reference levels and gamma dose rate criteria
Serbia	$AI_1 = \dfrac{C_{Ra}}{300} + \dfrac{C_{Th}}{200} + \dfrac{C_K}{3000}$ $AI_2 = \dfrac{C_{Ra}}{400} + \dfrac{C_{Th}}{300} + \dfrac{C_K}{5000}$ $AI_3 = \dfrac{C_{Ra}}{700} + \dfrac{C_{Th}}{500} + \dfrac{C_K}{8000}$	No	Bulk material $AI_1 \leq 1$ Bulk material for outside use $AI_2 \leq 1$ Bulk material for road construction $AI_3 \leq 1$	1 mSv/a
Slovakia	$AI = \dfrac{C_{Ra}}{300} + \dfrac{C_{Th}}{200} + \dfrac{C_K}{3000}$	Yes	Action level: $C_{Ra} \sim 120$ Bq/kg $AI \leq 1$ for bulk materials $AI \leq 2$ for materials with restricted use	An action level of 0.5 μSv/h
Slovenia		Yes	$C_{Ra} \leq 300$ Bq/kg $C_{Th} \leq 200$ Bq/kg $C_K \leq 5000$ Bq/kg	A limit of 0.1 μSv/h above background levels at a distance of 1 m from the surface

TABLE IV–1. SUMMARY OF THE LEGISLATION AND RECOMMENDATIONS USED IN SOME MEMBER STATES FOR THE REGULATORY CONTROL OF RADIONUCLIDES IN BUILDING AND CONSTRUCTION MATERIALS (cont.)

Country	Activity index	Regulation of Ra-226 content?	Decision values	Reference levels and gamma dose rate criteria
Tajikistan	$C_{eff} = C_{Ra} + 1.30\ C_{Th} + 0.09\ C_K$	No	$C_{eff} \leq 370\ \text{Bq/kg}$ If $C_{eff} > 4000\ \text{Bq/kg}$, the materials should not be used in building	
Türkiye	$AI = \dfrac{C_{Ra}}{300} + \dfrac{C_{Th}}{200} + \dfrac{C_K}{3000}$	No	$AI \leq 1$	
Ukraine	$C_{eff} = C_{Ra} + 1.31\ C_{Th} + 0.085\ C_K$	No	$C_{eff} \leq 370\ \text{Bq/kg}$	Mandatory levels: 30 microroentgen per hour for new dwellings and 50 microroentgen per hour for existing dwellings

[a] Exemption level for the use of phosphogypsum in agriculture or in the cement industry.

The following observations can be drawn from Table IV–1.

— The majority of the national regulations use the activity index approach. However, the formula for this index and the decision values vary from one State to another as a result of differences in the activity concentration in common building materials in each State and their final uses, variations in the background radiation and differences in the standard room model being used. Many States use 1 mSv per year as the reference level for building and construction materials. Armenia, the Czech Republic and Israel use 0.3 mSv per year above background as the reference level for building and construction materials. Finland and Slovenia use 0.1 mSv per year as the reference level for bulk and superficial materials used for road construction.
— In Argentina, there are no specific guidelines for the activity concentration of radionuclides in construction materials; however, the Basic Standard for Radiological Safety [IV–4] sets a reference level for the annual effective dose to the representative person of 1 mSv in existing exposure situations due to exposure to radionuclides in commodities such as building materials, food, feed and drinking water.
— Armenia, Belarus, Kazakhstan, the Republic of Moldova, the Russian Federation, Tajikistan and Ukraine use the ^{226}Ra equivalent activity concentration C_{eff}. Brazil established an exemption level for ^{226}Ra and ^{228}Ra for the use of phosphogypsum in agriculture or the cement industry.
— Albania and Finland include ^{137}Cs when calculating the activity concentration in construction materials due to the Chernobyl accident.
— Austria and Israel use activity indexes that directly take radon emanation into consideration. Conversely, countries including China, the Czech Republic, Latvia, Poland and Slovakia limit the ^{226}Ra activity concentration to account for radon exhalation. Brazil established an exemption level for the use of phosphogypsum in agriculture or in the cement industry, and a limiting value of 1000 Bq/kg is applied to the activity concentrations of ^{226}Ra or ^{228}Ra.
— Whether an activity index can be accurately calculated largely depends on the determination of activity concentrations. The legislation in Poland and the Russian Federation directly includes the measurement uncertainty in the determination of the activity index. Other States, including China, require that the relative measurement uncertainty should not exceed 20%.

More details on the regulations and method for implementation in some Member States are provided below.

IV–3. THE CZECH REPUBLIC

The legislation in the Czech Republic in relation to radionuclides in building and construction materials takes both gamma and radon exposure into consideration. The activity index AI is used as a screening tool. When the index AI of a material is higher than 0.5, which corresponds to a dose of 0.3 mSv per year, a cost–benefit analysis should be done by the producer of the material in question, considering that the public exposure should be reduced to a level as low as practically possible. The legislation also aims to control radon exhalation from building and construction materials; the activity concentration ^{226}Ra in these materials should not exceed the values listed in the in Table IV–2.

In the Czech Republic, the requirements for the measurement of radionuclide activity concentrations in building and construction materials are defined by the Atomic Act. Under this Act, manufacturers and importers of materials are required to ensure the activity concentrations of ^{226}Ra, and other relevant natural radionuclides are measured by qualified measurement service providers. There are currently approximately 12 laboratories in the Czech Republic that have permission from the State Office for Nuclear Safety to conduct such measurements. Manufacturers and importers of building and construction materials are also obliged to report the measurement results to the State Office for Nuclear Safety, and the manufacturers and importers make the measurement results available to the public on request.

If the activity concentrations of radionuclides in the tested building and construction materials exceed the permitted values set by the State Office for

TABLE IV–2. LIMIT VALUES FOR Ra-226 ACTIVITY CONCENTRATION IN CZECH LEGISLATION (NATURALLY OCCURRING RADIOACTIVE MATERIALS IN CONSTRUCTION)

Type of building material	Ra-226 limit value (Bq/kg)	
	Buildings where people live or stay	Other constructions where people do not live or stay
Material used in bulk amount (e.g. brick, concrete, gypsum)	150	500
Other material used in small amounts (e.g. tile) and raw material (e.g. sand, building stone, gravel aggregate, bottom ash)	300	1000

Nuclear Safety, the materials may not be permitted to enter the market. The State Office for Nuclear Safety in the Czech Republic published its Recommendations for the Measurement and Evaluation of the Content of Natural Radionuclides in Building Materials in November 2017 [IV–6]. This publication details the recommendations for sampling, treatment and measurement of samples, and related documentation, as well as approaches to evaluation measurement results. In addition, this publication provides guidance on situations where the sample measurement results exceed the permitted values.

The legislation in the Czech Republic does not require any measurement of the activity concentration of ^{226}Ra in nationally produced building and construction materials, and no limit value has been specified for such materials. It is the interest of builders and/or importers to avoid such materials with an activity concentration of ^{226}Ra exceeding the permitted values.

IV–4. FINLAND

In the Radiation Safety Guide [IV–7] entitled The Radioactivity of Building Materials and Ash, released by the Finnish Radiation and Nuclear Safety Authority (STUK) in 2010 [IV–7], four categories of building and construction materials are considered, following the approach described in the flow chart below. In addition to natural radionuclides such as ^{226}Ra, ^{232}Th and ^{40}K, the Finnish regulation also takes ^{137}Cs into account, since some industrial by-products also contain such radionuclides due to fallout from nuclear weapons tests and nuclear accidents (see Fig. IV–1).

The Radiation Safety Guide states that "a party running a radiation practice (responsible party) means any business or sole trader, enterprise, corporation or institution engaged in operations in which the exposure of human beings to natural radiation causes or is liable to cause a detriment to health". The responsible party is obligated to ensure that all the radiation safety requirements are satisfied in the life cycle, including production, use, handling and disposal of construction materials and ash. In addition, the responsible party is required to make certain that all necessary investigations and measurements to ensure safety are performed. The Radiation Safety Guide also requires the professional manufacturers, refiners, suppliers and end users of building materials to inform the next user regarding the radioactivity contained in the material [IV–7].

In practice, whoever places building and construction materials on the market in Finland is responsible for determining the exposure caused by the material in its intended use if the exposure could exceed the reference level. Specifically, the party placing the building and construction materials on the market is responsible for the following:

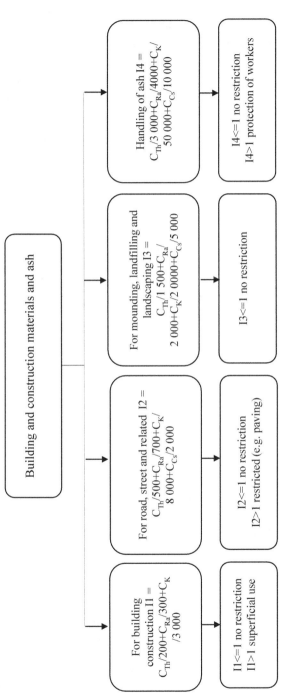

FIG. IV–1. Schematic presentation of regulatory requirements on building and construction materials in Finland. In this chart, C_{Ra}, C_{Th}, C_K and C_{Cs} denote the ^{226}Ra, ^{232}Th, ^{40}K and ^{137}Cs activity concentrations in a building and construction material (Bq/kg).

65

- Arranging for, and bearing the costs of, the sampling, measurements and dose assessments needed to determine the exposure caused by the material;
- Communicating the results of the determinations to STUK;
- Taking measures to reduce doses if the reference level is exceeded;
- Providing information to users of the building material on the levels of radioactivity in the material, and providing information on using the material in such a manner that the reference level will not be exceeded.

IV–5. CHINA

In the document entitled Limits of Radionuclides in Building Materials, released by the Standardization Administration of the People's Republic of China in 2010 (GB 6566-2010) [IV–8], both an internal exposure index (I_{Ra}) and external exposure index (I_r) are used to account for public exposure from building and construction materials. Commonly regulated construction materials include stone materials and ceramic tiles.

The internal exposure index is calculated as:

$$IRa = \frac{C_{Ra}}{200} \tag{IV–1}$$

where

C_{Ra} is the ^{226}Ra activity concentration in a building and construction material (Bq/kg);

and 200 Bq/kg is the maximum activity concentration for the ^{226}Ra activity concentration in a building and construction material.

The external exposure index is calculated as:

$$Ir = \frac{C_{Ra}}{370} + \frac{C_{Th}}{260} + \frac{C_K}{4200} \tag{IV–2}$$

where

C_{Ra}, C_{Th} and C_K are the ^{226}Ra, ^{232}Th and ^{40}K activity concentrations in a building and construction material (Bq/kg).

The maximum allowable activity concentration for ^{226}Ra is 370 Bq/kg if a construction material only contains ^{226}Ra. The maximum activity concentrations for ^{232}Th and ^{40}K allowed in a construction material are 260 Bq/kg and 4200 Bq/kg, respectively, if the construction material only contains ^{232}Th or ^{40}K.

Two random samples of no less than 2 kg each should be prepared for the measurement of ^{226}Ra, ^{232}Th and ^{40}K activity concentrations in the material. For example, for stone materials and ceramic tiles, a sample of no less than 2 kg per 5000 m^2 construction surface area needs to be submitted and measured. One of the samples should be sealed and kept aside, and the other one is chosen as the measurement sample.

The measurement sample of the test material is crushed so that the size of the grains does not exceed 0.16 mm. The activity concentrations of ^{226}Ra, ^{232}Th and ^{40}K in the test sample are measured using a low background multichannel gamma ray spectrometer.

When the sum of the activity concentrations of ^{226}Ra, ^{232}Th and ^{40}K in a test sample is greater than 37 Bq/kg, the measurement uncertainty (with a coverage factor of $k = 1$) should not exceed 20%.

The maximum allowed values of internal and external exposure indices are defined based on various categories of materials and buildings, as shown in Fig. IV–2.

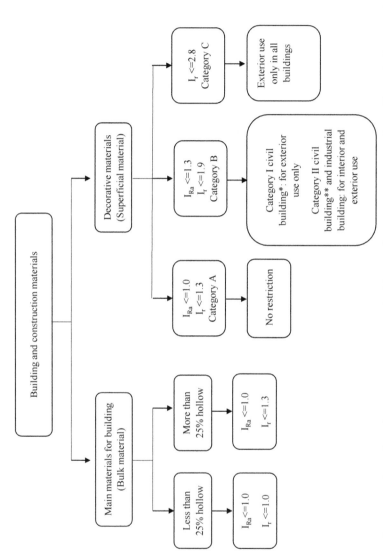

FIG. IV–2. *Schematic presentation of regulatory requirements on building and construction materials in China. * Category I civil building: residential dwelling, retirement residence, day care, hospital, school, office building, hotel, etc. ** Category II civil building: shopping mall, building for entertainment and cultural events, bookstore, library, exhibition hall, museum, stadium, waiting room for public transportation, restaurant, hair salon, etc.*

REFERENCES TO ANNEX IV

[IV–1]   EUROPEAN COMMISSION, FOOD AND AGRICULTURE ORGANIZATION OF THE UNITED NATIONS, INTERNATIONAL ATOMIC ENERGY AGENCY, INTERNATIONAL LABOUR ORGANIZATION, OECD NUCLEAR ENERGY AGENCY, PAN AMERICAN HEALTH ORGANIZATION, UNITED NATIONS ENVIRONMENT PROGRAMME, WORLD HEALTH ORGANIZATION, Radiation Protection and Safety of Radiation Sources: International Basic Safety Standards, IAEA Safety Standards Series No. GSR Part 3, IAEA, Vienna (2014).

[IV–2]   INTERNATIONAL ATOMIC ENERGY AGENCY, Status of Radon Related Activities in Member States Participating in Technical Cooperation Projects in Europe, IAEA-TECDOC-1810, IAEA, Vienna (2017).

[IV–3]   SCHROEYERS, W., Naturally Occurring Radioactive Materials in Construction: Integrating Radiation Protection in Reuse (COST Action Tu1301 NORM4BUILDING), Woodhead Publishing, Cambridge, UK (2017).

[IV–4]   ARGENTINA NUCLEAR REGULATORY AUTHORITY, Basic Radiological Safety Standard AR 10.1.1 Rev. 4, Paragraph 117, ARN, Buenos Aires (2019).

[IV–5]   BRAZILIAN NUCLEAR ENERGY COMMISSION, The Use of Phosphogypsum in the Agriculture and Cement Industries, CNEN Resolution 179/14 (2014).

[IV–6]   STATE OFFICE FOR NUCLEAR SAFETY, Recommendations for the Measurement and Evaluation of the Content of Natural Radionuclides in Building Materials, SÚJB/OS/18895/2017, SÚJB, Prague (2017).

[IV–7]   RADIATION AND NUCLEAR SAFETY AUTHORITY FINLAND (STUK), The Radioactivity of Building Materials and Ash, STUK, Helsinki (2010), https://www.finlex.fi/data/normit/23857-ST12-2e.pdf

[IV–8]   GENERAL ADMINISTRATION OF QUALITY SUPERVISION, INSPECTION AND QUARANTINE OF THE PEOPLE'S REPUBLIC OF CHINA, STANDARDIZATION ADMINISTRATION OF THE PEOPLE'S REPUBLIC OF CHINA, GB 6566-2010, Limits of radionuclides in building materials, AQSIQ, SAC, Beijing (2010).

CONTRIBUTORS TO DRAFTING AND REVIEW

Colgan, P.A.	International Atomic Energy Agency
Gehrcke, K.	Germany
German, O.	International Atomic Energy Agency
Lindner-Leschinski, E.M.	Austrian Agency for Health and Food Safety, Austria
Löfqvist, L.G.	Mark- och Miljökontroll i Särö AB, Sweden
Zhou, L.G.	National Research Council Canada, Canada

Consultants Meetings

Vienna, Austria: 6–9 May 2019, 14–17 October 2019,
17–21 August 2020, 23–24 February 2021

ORDERING LOCALLY

IAEA priced publications may be purchased from the sources listed below or from major local booksellers.

Orders for unpriced publications should be made directly to the IAEA. The contact details are given at the end of this list.

NORTH AMERICA

Bernan / Rowman & Littlefield
15250 NBN Way, Blue Ridge Summit, PA 17214, USA
Telephone: +1 800 462 6420 • Fax: +1 800 338 4550
Email: orders@rowman.com • Web site: www.rowman.com/bernan

REST OF WORLD

Please contact your preferred local supplier, or our lead distributor:

Eurospan Group
Gray's Inn House
127 Clerkenwell Road
London EC1R 5DB
United Kingdom

Trade orders and enquiries:
Telephone: +44 (0)176 760 4972 • Fax: +44 (0)176 760 1640
Email: eurospan@turpin-distribution.com

Individual orders:
www.eurospanbookstore.com/iaea

For further information:
Telephone: +44 (0)207 240 0856 • Fax: +44 (0)207 379 0609
Email: info@eurospangroup.com • Web site: www.eurospangroup.com

Orders for both priced and unpriced publications may be addressed directly to:
Marketing and Sales Unit
International Atomic Energy Agency
Vienna International Centre, PO Box 100, 1400 Vienna, Austria
Telephone: +43 1 2600 22529 or 22530 • Fax: +43 1 26007 22529
Email: sales.publications@iaea.org • Web site: www.iaea.org/publications